Lecture Notes in Mathematics

2114

T0236549

More information about this series at
http://www.springer.com/series/304

Elena Shchepakina • Vladimir Sobolev •
Michael P. Mortell

Singular Perturbations

Introduction to System Order Reduction
Methods with Applications

Springer

Elena Shchepakina
Vladimir Sobolev
Samara State Aerospace University
Samara
Russia

Michael P. Mortell
University College Cork
Cork
Ireland

ISBN 978-3-319-09569-1   ISBN 978-3-319-09570-7 (eBook)
DOI 10.1007/978-3-319-09570-7
Springer Cham Heidelberg New York Dordrecht London

Lecture Notes in Mathematics ISSN print edition: 0075-8434
ISSN electronic edition: 1617-9692

Library of Congress Control Number: 2014950898

Mathematics Subject Classification (2014): 34-02, 34C45, 34D15, 34E15, 34E17, 37D10, 70K70, 80M35, 92C45, 93C70

Printed on acid-free paper

Springer is part of Springer Science+Business Media (www.springer.com)

We dedicate this book to the memory of our friend and colleague Alexei Pokrovskii (1948–2010)

Foreword

This book is a tribute to Alexei Pokrovskii (1948–2010) who introduced the Russian and Irish coauthors and who generously promoted and made important contributions to our understanding of singular perturbations from an applied and geometric perspective.

The three authors continue and explain many recent results on the asymptotics of slow integral (or invariant) manifolds and their stability. They do so by cleverly describing a series of illustrative examples of increasing complexity and reality. Many applications in chemical kinetics are particularly impressive, as is their ultimate study of two-dimensional canards and higher-dimensional black swans.

Readers of this clear and well-motivated monograph will be prepared to advance to the even more sophisticated research literature that takes a dynamical systems approach to multi scale systems. The authors are to be congratulated on completing a tough job, very well done. They've certainly earned our appreciation and that of future students.

Seattle, WA   Robert O'Malley
December 2013

Preface

The idea of using a small parameter to set up a perturbation series has been with us since at least the work of Stokes in 1847[1] on the investigation of water waves. The use of integral manifolds, with a small parameter, is of more recent vintage. It can be found in [45, 72, 75, 92, 114, 170, 197, 217, 218]. Over the past 50 years there have been many books devoted to regular and singular perturbations, but there are few books in which singular perturbations are combined with integral manifolds. Moreover, many of these were published only in Russian. The purpose of the present book is to fill this gap.

We deal with a system of first order ODEs some of which are singularly perturbed, i.e., when the small parameter is set to zero the ability to satisfy all initial conditions is lost. We introduce a method for the qualitative analysis of these singularly perturbed ODEs. The method relies on the theory of integral manifolds, which essentially replaces the original system by another system on an integral manifold of lower dimension. The lowering of the dimension occurs due to the decomposition of the original system in the vicinity of the integral surface into the independent "slow" subsystem and the "fast" subsystem. If the slow integral manifold is attracting, then the analysis of the original system can be replaced by the analysis of the slow subsystem. In the language of perturbation theory a slow integral manifold is associated with the outer (slow) solution and a fast integral manifold is associated with boundary layer (fast) corrections.

The book proceeds with the interplay of theory and illustrative examples, in many cases taken from physical problems. There are many such examples in Chap. 3, where the reader is introduced at an easy pace to the use of the theory. As the chapters progress, the theory and corresponding examples become more sophisticated. In Chaps. 7 and 8 we deal with systems where the usual hypotheses in integral manifold theory are violated. The method of solution is then illustrated

---

[1]G.G. Stokes, On the theory of oscillatory waves. Camb Trans 8:441–473.

by a series of examples on gyroscopic motion, control problems, and a model of thermal explosion. These problems can be quite difficult, so much of the detailed calculation is given. In Chap. 8 the concepts of canard and black swan are introduced and illustrated by examples on the van der Pol oscillator, a fast phages–slow bacteria model, and some laser and chemical models. There is also a detailed discussion of two classical combustion models, including the calculation of the critical value of the parameter that separates explosive from non-explosive regimes. In Chap. 9 the proofs of certain theorems are given that have been signalled earlier in the book. These require a more mature reader.

    The authors are grateful to Robert O'Malley who was there at the beginning and gave much valuable advice, as well as Grigory Barenblatt, Eric Benoit and Jean Mawhin for helpful discussions. This work is supported in part by the Russian Foundation for Basic Research (grants 12-08-00069, 13-01-97002, 14-01-97018, 14-08-91373), TUBITAK (grant 113E595), Division on the EMMCP of Russian Academy of Sciences, Program for basic research no. 14, project 1.12, and the Ministry of Education and Science of the Russian Federation in the framework of the implementation of Program of increasing the competitiveness of SSAU for 2013–2020 years.

Cork, Ireland   Michael P. Mortell
Samara, Russia   Elena Shchepakina
Samara, Russia   Vladimir Sobolev
January 2014

Contents

Chapter 1
Introduction

Abstract Chapter 1 provides an easy introduction to perturbation methods. It begins with an algebraic equation and proceeds to a second order ODE. The concept of an initial or boundary layer is introduced. This motivates the method of multiple scales. The idea of slow surfaces and slow integral manifolds is introduced and illustrative examples are given. Then a statement of Tikhonov's theorem is given which answers the question about the permissibility of the application of a "degenerate" system ($\varepsilon = 0$) as a zero-approximation to the full system.

1.1 Regular and Singular Perturbations

1.1.1 Algebraic Equations

In this subsection we introduce the basic ideas of perturbation theory as applied to the problem of finding roots of polynomials.

By way of illustration, we consider two quadratic equations

$$\mu^2 + b\mu + \varepsilon a = 0, \tag{1.1}$$

with roots μ_1, μ_2 given by

$$\mu_1 = \mu_1(\varepsilon) = \frac{-b + \sqrt{b^2 - 4\varepsilon a}}{2}, \quad \mu_2 = \mu_2(\varepsilon) = \frac{-b - \sqrt{b^2 - 4\varepsilon a}}{2}, \tag{1.2}$$

and

$$\varepsilon \lambda^2 + b\lambda + a = 0, \tag{1.3}$$

with roots λ_1, λ_2 given by

$$\lambda_1 = \lambda_1(\varepsilon) = \frac{-b + \sqrt{b^2 - 4\varepsilon a}}{2\varepsilon}, \quad \lambda_2 = \lambda_2(\varepsilon) = \frac{-b - \sqrt{b^2 - 4\varepsilon a}}{2\varepsilon}. \tag{1.4}$$

© Springer International Publishing Switzerland 2014
E. Shchepakina et al., *Singular Perturbations*, Lecture Notes in Mathematics 2114,
DOI 10.1007/978-3-319-09570-7_1

Here a and b are given constants, and $\varepsilon > 0$ is a small parameter. For the sake of definiteness we suppose b positive and a nonnegative.

We say that (1.1) is *regularly perturbed*, and (1.3) is *singularly perturbed* when $\varepsilon \to 0$. To clarify these ideas, we begin by setting $\varepsilon = 0$ in (1.1) and (1.3). Thus we get

$$\mu^2 + b\mu = 0,$$

with roots $\bar{\mu}_1 = 0$ and $\bar{\mu}_2 = -b$, and

$$b\lambda + a = 0,$$

with the single root $\bar{\lambda}_1 = -a/b$.

It is easy to see that as $\varepsilon \to 0$ in (1.1)

$$\mu_1(\varepsilon) \to \bar{\mu}_1 = 0, \quad \text{and} \quad \mu_2(\varepsilon) \to \bar{\mu}_2 = -b.$$

However, although

$$\lambda_1(\varepsilon) \to \bar{\lambda}_1 = -a/b \quad \text{as} \quad \varepsilon \to 0,$$

the root corresponding to $\lambda_2(\varepsilon)$ is lost. This loss of a root follows from the fact that when we set $\varepsilon = 0$ in (1.3), the resulting equation is reduced in order from a quadratic to a linear equation. From (1.2) and (1.4) it follows that $\mu_1(\varepsilon), \mu_2(\varepsilon)$ are continuous at $\varepsilon = 0$, but $\lambda_2(\varepsilon)$ is not.

Thus, the distinguishing features of *singular* perturbations are the reduction in order of the original equation and the loss of continuity for some solution(s) at $\varepsilon = 0$.

We note that the change of variable $\lambda = \mu/\varepsilon$ reduces the singular equation (1.3) to the regular equation (1.1). Changes of variables of this type are typical of perturbation theory, and it is standard practice to seek such transformations.

Taylor expansions for the roots $\mu_i, \lambda_i, \ i = 1, 2$, given by (1.2) and (1.4) yield the result, for small ε,

$$\mu_1(\varepsilon) = -\varepsilon a/b - \varepsilon^2 a^2/b^3 + \dots, \tag{1.5}$$

and

$$\mu_2(\varepsilon) = -b + \varepsilon a/b + \varepsilon^2 a^2/b^3 + \dots, \tag{1.6}$$

with

$$\lambda_{1,2}(\varepsilon) = \mu_{1,2}(\varepsilon)/\varepsilon. \tag{1.7}$$

The question we now wish to address is whether we can recover the expansions (1.5)–(1.7) directly from the original equations (1.1) and (1.3) without recourse to the exact solutions (1.2) and (1.4). To this end, the notion of an asymptotic expansion is introduced.

1.1.2 Asymptotic Expansions

The search for the roots of polynomial equations is probably the simplest problem in the theory of perturbations, and we use Eqs. (1.1) and (1.3) to illustrate the idea of an asymptotic expansion. We use the usual order symbols O and o that are defined as follows. Let $\phi(\varepsilon)$ and $\psi(\varepsilon)$ be given functions with

$$\left| \frac{\phi(\varepsilon)}{\psi(\varepsilon)} \right| \leq C < \infty.$$

Then we write

$$\phi(\varepsilon) = O(\psi(\varepsilon)) \text{ as } \varepsilon \to 0.$$

If

$$\frac{\phi(\varepsilon)}{\psi(\varepsilon)} \to 0 \text{ as } \varepsilon \to 0,$$

we write

$$\phi(\varepsilon) = o(\psi(\varepsilon)) \text{ as } \varepsilon \to 0.$$

To find an asymptotic expansion to approximate the roots of (1.1), we write the formal series, called a regular perturbation expansion,

$$\mu(\varepsilon) = \alpha_0 + \varepsilon \alpha_1 + \varepsilon^2 \alpha_2 + O(\varepsilon^3). \tag{1.8}$$

On substituting (1.8) into (1.1) and equating to zero the coefficients of the various powers of ε, we find the sequence of equations for $\alpha_0, \alpha_1, \alpha_2, \ldots$

$$\alpha_0^2 + b\alpha_0 = 0,$$

$$(2b\alpha_0 + b)\alpha_1 + a = 0,$$

$$(2b\alpha_0 + b)\alpha_2 + (\alpha_1)^2 = 0,$$

$$\cdots .$$

Then

$$\alpha_0 = 0 \text{ or } \alpha_0 = -b.$$

For $\alpha_0 = 0$, $\alpha_1 = -a/b$ and $\alpha_2 = -a^2/b^3$, and (1.5) is recovered.
For $\alpha_0 = -b$, $\alpha_1 = a/b$ and $\alpha_2 = a^2/b^3$, and (1.6) is recovered.

Thus, the expansions (1.5) and (1.6) for $\mu_1(\varepsilon)$ and $\mu_2(\varepsilon)$ are found unless we have the intuition, or a procedure to tell us, that the asymptotic expansion for the roots $\lambda_1(\varepsilon), \lambda_2(\varepsilon)$ of (1.3) takes the form

$$\lambda(\varepsilon) = \gamma_0/\varepsilon + \gamma_1 + \varepsilon\gamma_2 + O(\varepsilon^2),$$

that yields $\gamma_0 = 0$ or $\gamma_0 = -b$ on substitution into (1.3). However, recognising that the transformation $\lambda = \mu/\varepsilon$ reduces (1.3) to (1.1), the expansions for $\lambda_{1,2}(\varepsilon)$ can again be found by a regular perturbation.

We now write the asymptotic expansion for $\mu_{1,2}(\varepsilon)$ as

$$\mu_1(\varepsilon) = -\varepsilon a/b - \varepsilon^2 a^2/b^3 + O(\varepsilon^3),$$

$$\mu_2(\varepsilon) = -b + \varepsilon a/b + \varepsilon^2 a^2/b^3 + O(\varepsilon^3),$$

and for $\lambda_{1,2}(\varepsilon)$ as

$$\lambda_1(\varepsilon) = -a/b - \varepsilon a^2/b^3 + O(\varepsilon^2),$$

$$\lambda_2(\varepsilon) = -b/\varepsilon + a/b + \varepsilon a^2/b^3 + O(\varepsilon^2).$$

1.1.3   Second Order Differential Equation

As the next introductory example, we consider the second order ordinary differential equation

$$\varepsilon\ddot{x} + b\dot{x} + ax = 0; \quad x(0) = x_0, \quad \dot{x}(0) = y_0, \tag{1.9}$$

where the dot refers to differentiation with respect to time t. Equation (1.9) describes the motion of a damped spring with "small" mass ε, $0 < \varepsilon \ll 1$, if $b > 0$ and $a > 0$ and both are constants of $O(1)$. The initial extension of the spring is x_0 and the initial velocity of the mass is y_0, both of which are given.

This is a linear homogeneous ordinary differential equation with constant coefficients, and is solved by seeking solution of the form $x = e^{\lambda t}$. Then λ satisfies

$$\varepsilon\lambda^2 + b\lambda + a = 0,$$

with solutions $\lambda_{1,2}(\varepsilon)$ given by (1.4). The solution to the problem (1.9) then is

$$x(t) = C_1 e^{\lambda_1(\varepsilon)t} + C_2 e^{\lambda_2(\varepsilon)t} \tag{1.10}$$

where the constants C_1 and C_2 are determined by the initial conditions to be

$$C_1 = \frac{\lambda_2 x_0 - y_0}{\lambda_2 - \lambda_1}, \quad C_2 = \frac{y_0 - \lambda_1 x_0}{\lambda_2 - \lambda_1}. \tag{1.11}$$

From the form of $\lambda_{1,2}(\varepsilon)$, we note that if $b^2 > 4a\varepsilon$ the motion of the spring is critically damped, i.e., there are no oscillations. It is seen from (1.4) that

$$\lambda_1(\varepsilon) = -\frac{a}{b} + O(\varepsilon), \text{ and } \lambda_2(\varepsilon) = -\frac{b}{\varepsilon} + O(1),$$

and the dominant terms in the solution $x(t, \varepsilon)$ are

$$x(t, \varepsilon) = C_1 e^{-(\frac{a}{b} + O(\varepsilon))t} + C_2 e^{-(\frac{b}{\varepsilon} + O(1))t}.$$

Since $0 < \varepsilon \ll 1$, there is a "fast" decay associated with the C_2 term as compared with the slow decay of the C_1 term. Thus the term $C_2 e^{\lambda_2(\varepsilon)t}$ is of significance only near the initial point $t = 0$.

This observation motivates the introduction of an "initial layer" or "boundary layer", where the natural independent variable is the "stretched" variable $\tau = t/\varepsilon$. The independent variable t is then associated with the "outer region" at a remove from $t = 0$. We will construct an asymptotic expansion for the solution $x(t, \varepsilon)$ involving the variables t and τ.

Following [127] or [205], *an outer expansion*

$$x_{\text{outer}}(t, \varepsilon) = f_0(t) + \varepsilon f_1(t) + O(\varepsilon^2), \tag{1.12}$$

and *an initial layer* expansion

$$x_{\text{layer}}(\tau, \varepsilon) = g_0(\tau) + \varepsilon g_1(\tau) + O(\varepsilon^2) \tag{1.13}$$

are assumed, where $\tau = t/\varepsilon$. The condition

$$x_{\text{layer}}(\tau, \varepsilon) \to 0, \text{ as } \tau \to +\infty \tag{1.14}$$

is also imposed.

On substitution the outer expansion (1.12) into the Eq. (1.9) and equating powers of ε, the equations for $f_0(t)$ and $f_1(t)$ are

$$b\frac{df_0(t)}{dt} + af_0(t) = 0,$$

$$b\frac{df_1(t)}{dt} + af_1(t) = -\frac{d^2 f_0(t)}{dt^2}.$$

Then

$$f_0(t) = c_1 e^{-at/b}$$

and

$$f_1(t) = c_2 e^{-at/b} - t \frac{a^2}{b^3} c_1 e^{-at/b},$$

where the constants c_1, c_2 remain to be determined. Then

$$x_{\text{outer}}(t, \varepsilon) = c_1 e^{-at/b} + \varepsilon \left[c_2 e^{-at/b} - t \frac{a^2}{b^3} c_1 e^{-at/b} \right] + O(\varepsilon^2).$$

On using the variable $\tau = t/\varepsilon$, Eq. (1.9) becomes

$$\frac{d^2 x}{d\tau^2} + b \frac{dx}{d\tau} + \varepsilon a x = 0. \tag{1.15}$$

Substituting the expansion (1.13) into Eq. (1.15), equating powers of ε and noting the condition (1.14), yields

$$\frac{d^2 g_0}{d\tau^2} + b \frac{dg_0}{d\tau} = 0, \quad g_0(\infty) = 0,$$

$$\frac{d^2 g_1}{d\tau^2} + b \frac{dg_1}{d\tau} = -a g_0, \quad g_1(\infty) = 0.$$

Then,

$$g_0 = d_1 e^{-b\tau}$$

and

$$g_1 = d_2 e^{-b\tau} - \frac{a}{b} \tau d_1 e^{-b\tau},$$

where the constants d_1, d_2 remain to be determined. Then

$$x_{\text{layer}}(\tau, \varepsilon) = d_1 e^{-b\tau} + \varepsilon \left[d_2 e^{-b\tau} - \frac{a}{b} \tau d_1 e^{-b\tau} \right] + O(\varepsilon^2).$$

We now assume that the solution $x(t, \varepsilon)$ of (1.9) is of the form

$$x(t, \varepsilon) = x_{\text{outer}}(t, \varepsilon) + x_{\text{layer}}(\tau, \varepsilon).$$

The initial conditions associated with Eq. (1.9), with the definitions (1.12) and (1.13), imply

$$f_0(0) + g_0(0) = x_0, \quad g_0'(0) = 0,$$

and

$$f_1(0) + g_1(0) = 0, \quad f_0'(0) + g_1'(0) = y_0,$$

on recalling that $\frac{d}{dt} = \frac{1}{\varepsilon}\frac{d}{d\tau}$, f', g' refer to differentiation with respect to t and τ respectively. Then $d_1 = 0$ and $c_1 = x_0$, while $d_2 = -c_2$ and $c_2 = \frac{1}{b}(y_0 + \frac{a}{b}x_0)$.

The asymptotic expansion is

$$x(t, \varepsilon) = x_0 e^{-at/b} + \varepsilon \left[\left(\frac{y_0}{b} + \frac{ax_0}{b^2} \right) - t \frac{a^2}{b^3} x_0 \right] e^{-at/b}$$

$$-\varepsilon \left(\frac{y_0}{b} + \frac{ax_0}{b^2} \right) e^{-bt/\varepsilon} + O(\varepsilon^2). \tag{1.16}$$

On examining the exact solution (1.10) and (1.11), it is clear, on noting the asymptotic forms of $\lambda_{1,2}(\varepsilon)$, that

$$x_{\text{outer}}(t, \varepsilon) = \frac{\lambda_2 x_0 - y_0}{\lambda_2 - \lambda_1} e^{\lambda_1 t}$$

and

$$x_{\text{layer}}(\tau, \varepsilon) = \frac{y_0 - \lambda_1 x_0}{\lambda_2 - \lambda_1} e^{\lambda_2 t},$$

and the validity of the expansion (1.16) is easily checked. It should be noted that the expansion (1.16) satisfies the displacement condition to $O(\varepsilon)$ and the velocity condition to $O(1)$. Furthermore, the initial conditions are applied to the full solution $x(t, \varepsilon)$ and not just to $x_{\text{layer}}(\tau, \varepsilon)$, as would be in the case for the method of matched asymptotic expansions. The consequence is that the "matching" of the outer to the layer solution to determine the arbitrary constants is obviated.

1.2 Method of Multiple Scales

We have noted that the solution (1.10) and (1.11) of Eq. (1.9) has a 'fast' time scale $\tau = t/\varepsilon$ and a 'slow' time scale t. This observation can be exploited as follows.

1.2.1 Second Order Differential Equation

Equation (1.9) is rewritten as in (1.15), and we assume the solution of (1.15) in the form

$$x(\tau;\varepsilon) = F(\tau,t;\varepsilon), \quad t = \varepsilon\tau, \tag{1.17}$$

and regard t, τ as independent variables. In these variables, the initial conditions become

$$x(0;\varepsilon) = x_0 = F(0,0;\varepsilon) \tag{1.18}$$

and

$$\varepsilon\frac{dx}{dt}(0;\varepsilon) = \varepsilon y_0 = \frac{\partial F}{\partial \tau}(0,0;\varepsilon) + \varepsilon\frac{\partial F}{\partial t}(0,0;\varepsilon). \tag{1.19}$$

The asymptotic expansion for $x(\tau;\varepsilon)$ is assumed to be

$$F(\tau,t;\varepsilon) = F_0(\tau,t) + \varepsilon F_1(\tau,t) + \varepsilon^2 F_2(\tau,t) + \dots. \tag{1.20}$$

On substituting (1.20) into (1.15), equating powers of ε, and using the initial conditions (1.18) and (1.19), we get the following sequence of problems.

ε^0 :

$$\frac{\partial^2 F_0}{\partial \tau^2} + b\frac{\partial F_0}{\partial \tau} = 0,$$

$$F_0(0,0) = x_0, \quad \frac{\partial F_0}{\partial \tau}(0,0) = 0;$$

ε^1 :

$$\frac{\partial^2 F_1}{\partial \tau^2} + b\frac{\partial F_1}{\partial \tau} = -a F_0 - 2\frac{\partial^2 F_0}{\partial \tau \partial t} - b\frac{\partial F_0}{\partial t},$$

$$F_1(0,0) = 0, \quad \frac{\partial F_1}{\partial \tau}(0,0) = y_0 - \frac{\partial F_0}{\partial t}(0,0);$$

ε^2 :

$$\frac{\partial^2 F_2}{\partial \tau^2} + b\frac{\partial F_2}{\partial \tau} = -a F_1 - 2\frac{\partial^2 F_1}{\partial \tau \partial t} - b\frac{\partial F_1}{\partial t} - \frac{\partial^2 F_0}{\partial t^2},$$

$$F_2(0,0) = 0, \quad \frac{\partial F_2}{\partial \tau}(0,0) = -\frac{\partial F_1}{\partial t}(0,0).$$

Then

$$F_0(\tau, t) = A_0(t)e^{-b\tau} + B_0(t),$$

where, using the initial conditions,

$$A_0(0) = 0, \quad B_0(0) = x_0.$$

The right hand side at ε^1 is

$$- \left[\left(a A_0(t) - b A_0'(t) \right) e^{-b\tau} + a B_0(t) + b B_0'(t) \right].$$

In order to avoid any growth (secular terms) in the fast variable τ in $F_1(\tau, t)$, the r.h.s. is set to zero.

Thus

$$A_0'(t) - \frac{a}{b} A_0(t) = 0, \quad A_0(0) = 0,$$

$$B_0'(t) + \frac{a}{b} B_0(t) = 0, \quad B_0(0) = x_0,$$

and then

$$A_0(t) \equiv 0, \quad B_0(t) = x_0 e^{-\frac{a}{b}t}.$$

This gives

$$F_0(\tau, t) = x_0 e^{-\frac{a}{b}t}$$

and

$$F_1(\tau, t) = A_1(t)e^{-b\tau} + B_1(t).$$

The right hand side at ε^2 is

$$\left(b A_1' - a A_1 \right) e^{-b\tau} - \left(b B_1' + a B_1 \right) - x_0 \frac{a^2}{b^2} e^{-\frac{a}{b}t}.$$

Again we avoid secular terms in τ by setting

$$A_1' - \frac{a}{b} A_1 = 0$$

and

$$B_1' + \frac{a}{b} B_1 = -x_0 \frac{a^2}{b^2} e^{-\frac{a}{b}t},$$

where the initial conditions at this order yield

$$B_1(0) = -A_1(0) = \frac{1}{b}(y_0 + \frac{a}{b}x_0).$$

Then

$$B_1(t) = \frac{1}{b}(y_0 + \frac{a}{b}x_0)e^{-\frac{a}{b}t} - x_0\frac{a^2}{b^3}te^{-\frac{a}{b}t}$$

and

$$A_1(t) = -\frac{1}{b}(y_0 + \frac{a}{b}x_0)e^{\frac{a}{b}t}.$$

Finally,

$$x(\tau;\varepsilon) = F_0(\tau,t) + \varepsilon F_1(\tau,t) + O(\varepsilon^2)$$

$$= x_0e^{-\frac{a}{b}t} + \frac{\varepsilon}{b}(y_0 + \frac{a}{b}x_0)e^{-\frac{a}{b}t} - \varepsilon x_0\frac{a^2}{b^3}te^{-\frac{a}{b}t}$$

$$-\frac{\varepsilon}{b}(y_0 + \frac{a}{b}x_0)e^{\frac{a}{b}t}e^{-b\tau} + O(\varepsilon^2). \tag{1.21}$$

We note that the term $e^{\frac{a}{b}t}e^{-b\tau}$ can be replaced by $e^{-b\tau}$ due to the dominance of $e^{-b\tau}$ for $\tau > 0$, since

$$e^{\frac{a}{b}t}e^{-b\tau} = e^{-\frac{bt}{\varepsilon}(1-\varepsilon\frac{a}{b^2})},$$

and (1.21) agrees with (1.16).

1.2.2 Second Order Differential System

The second order differential equation discussed above can be rewritten as the planar differential system

$$\dot{x} = y, \ x(0) = x_0;$$
$$\varepsilon\dot{y} = -ax - by, \ y(0) = y_0. \tag{1.22}$$

The solution to (1.22) can be represented in the form

$$x = x_{\text{outer}}(t,\varepsilon) + x_{\text{layer}}(\tau,\varepsilon) \tag{1.23}$$

$$y = y_{\text{outer}}(t,\varepsilon) + y_{\text{layer}}(\tau,\varepsilon) \tag{1.24}$$

where x_{outer} and x_{layer} were defined above after (1.16), and

$$y_{\text{outer}}(t,\varepsilon) = \lambda_1 \frac{\lambda_2 x_0 - y_0}{\lambda_2 - \lambda_1} e^{\lambda_1 t} = \lambda_1 x_{\text{outer}}(t,\varepsilon), \tag{1.25}$$

$$y_{\text{layer}}(\tau,\varepsilon) = \lambda_2 \frac{y_0 - \lambda_1 x_0}{\lambda_2 - \lambda_1} e^{\lambda_2 t} = \lambda_2 x_{\text{layer}}(\tau,\varepsilon), \tag{1.26}$$

from the first of (1.22). It should be emphasized that

$$x = x_{\text{outer}}, \quad y = y_{\text{outer}},$$

and

$$x = x_{\text{layer}}, \quad y = y_{\text{layer}}$$

are exact solutions of the system of differential equations in (1.22), but do not satisfy the initial conditions.

We will use this system to illustrate some ideas in the next Section.

1.2.3   A Note on the Initial Conditions

Recall, that

$$x_{\text{outer}}(t,\varepsilon) = \frac{\lambda_2 x_0 - y_0}{\lambda_2 - \lambda_1} e^{\lambda_1 t}.$$

Thus, $x_{\text{outer}}(t,\varepsilon)$ satisfies the initial condition

$$x_{\text{outer}}(0,\varepsilon) = \frac{\lambda_2 x_0 - y_0}{\lambda_2 - \lambda_1},$$

which can be rewritten as follows

$$x_{\text{outer}}(0,\varepsilon) = x_0 + \frac{\lambda_2 x_0 - y_0}{\lambda_2 - \lambda_1} - x_0 = x_0 + \frac{\lambda_1 x_0 - y_0}{\lambda_2 - \lambda_1} = x_0 + \varepsilon \frac{y_0 - \lambda_1 x_0}{\sqrt{b^2 - 4\varepsilon a}}.$$

This means that the initial value of $x_{\text{outer}}(t,\varepsilon)$ is different from x_0 by the amount

$$\varepsilon \frac{y_0 - \lambda_1 x_0}{\sqrt{b^2 - 4\varepsilon a}} = O(\varepsilon).$$

1.2.4   A Note on the Meaning of "Small"

The equation for a mass (m), spring (k), dash-pot (v) system is

$$m\frac{d^2x}{dt^2} + v\frac{dx}{dt} + kx = 0,$$

where $x(t)$ is the displacement of the mass from its equilibrium point $x = 0$. By comparing terms in the equation, the dimensions of v are mT^{-1}, and of k are mT^{-2} where T is a measure of time. If we introduce a length scale x_0 and a time scale T_0, the dimensionless displacement and time are $\bar{x} = x/x_0$, $\bar{t} = t/T_0$, and the original equations becomes

$$\frac{m}{\bar{k}T_0^2}\frac{d^2\bar{x}}{d\bar{t}^2} + \frac{v}{\bar{k}T_0}\frac{d\bar{x}}{d\bar{t}} + a\bar{x} = 0,$$

where $k = a\bar{k}$ and a is a number.

We write the equation as

$$\varepsilon\frac{d^2\bar{x}}{d\bar{t}^2} + b\frac{d\bar{x}}{d\bar{t}} + a\bar{x} = 0,$$

where ε, a, b are dimensionless parameters. The meaning of a "small" mass then is defined by $0 < \varepsilon \ll 1$, where $\varepsilon = \dfrac{m}{\bar{k}T_0^2}$ is a dimensionless constant.

An essential task before using a perturbation scheme is to nondimensionalise the equations, and identify the small (large) parameter.

Note that prerequisites are an introductory course in perturbation methods e.g. parts of [6, 9, 17, 25, 28, 41, 62, 69, 76, 77, 80, 85–88, 95, 121, 122, 127, 153, 169, 203].

1.3   Singularly Perturbed Differential Systems

1.3.1   Slow Surfaces and Slow Integral Manifolds

Consider the system of ordinary differential equations

$$\frac{dx}{dt} = f(x, y, t, \varepsilon),$$

$$\varepsilon\frac{dy}{dt} = g(x, y, t, \varepsilon),$$

(1.27)

with $x \in \mathbb{R}^m$, $y \in \mathbb{R}^n$, $t \in \mathbb{R}$, and ε is a small positive parameter. Such systems are called *singularly perturbed systems*, since when $\varepsilon = 0$ the ability to specify an arbitrary initial condition for $y(t)$ is lost. The usual approach to the qualitative study of (1.27) is to consider first *the degenerate* system ($\varepsilon = 0$)

$$\frac{dx}{dt} = f(x, y, t, 0),$$
$$0 = g(x, y, t, 0), \tag{1.28}$$

and then to draw conclusions about the qualitative behavior of the full system (1.27) for sufficiently small ε.

In order to recall a basic result of the theory of singularly perturbed systems we introduce the following terminology and assumptions.
The system of equations

$$\frac{dx}{dt} = f(x, y, t, \varepsilon) \tag{1.29}$$

is called the slow subsystem of (1.27), x is called the slow variable and the system of equations

$$\varepsilon \frac{dy}{dt} = g(x, y, t, \varepsilon) \tag{1.30}$$

is called the fast subsystem of (1.27). Here $x \in \mathbb{R}^m$, $y \in \mathbb{R}^n$, $t \in \mathbb{R}$.

In this book we introduce a method for the qualitative asymptotic analysis of singularly perturbed ordinary differential equations. The method relies on the theory of integral manifolds, which essentially replaces the original system by another system on an integral manifold whose dimension is equal to that of the slow subsystem.

Definition 1. A smooth surface S in $\mathbb{R} \times \mathbb{R}^m \times \mathbb{R}^n$ is called an integral manifold of the system (1.27) if any integral curve of the system that has at least one point in common with S lies entirely on S.

Formally, if $(t_0, x(t_0), y(t_0)) \in S$, then the integral curve $(t, x(t, \varepsilon), y(t, \varepsilon))$ lies entirely on S. The only integral manifolds of system (1.27) discussed here are those of dimension m (the dimension of the slow variable x) that can be represented as graphs of vector-valued functions

$$y = h(x, t, \varepsilon).$$

Here it is assumed that $h(x, t, \varepsilon)$ is a sufficiently smooth function of ε. Such integral manifolds are called manifolds of slow motions — the origin of this term lies in nonlinear mechanics.

Definition 2. The surface described by the equation

$$g(x, y, t, 0) = 0$$

is called a slow surface. When the dimension of this surface is equal to one, it is called a slow curve.

So $\frac{dx}{dt} = f(x, y, t, \varepsilon)$ is the slow subsystem, while $g(x, y, t, 0) = 0$ is the slow surface. We also stipulate that $h(x, t, 0) = \phi(x, t)$, where $\phi(x, t)$ is a function whose graph is a sheet of the slow surface

$$g(x, y, t, 0) = 0,$$

i.e. the slow surface can be considered as a zero-order approximation of the slow integral manifold.

To explain the sense of this term (slow surface), it is enough to notice that the derivative of a fast variable y along a slow surface has small values; that is, the fast variable near to this surface changes slowly like the slow variable x:

$$\frac{dy}{dt} = \frac{g(x, y, t, \varepsilon)}{\varepsilon} = g_\varepsilon(x, y, t, 0) + O(\varepsilon),$$

since $g(x, y, t, 0) = 0$ on the slow surface.

It is also assumed that the equation $g(x, y, t, 0) = 0$ has an isolated root $\phi(x, t)$:

$$g(x, \phi(x, t), t, 0) \equiv 0.$$

Before we pursue the idea of integral manifolds of a differential equation we first consider several examples of the notation of sheets (or branches) of a surface.

As a first example, let x and y be scalar variables and $g = y^2 + x$. Then the equation

$$g = y^2 + x = 0$$

gives two roots $y = \phi(x) = -\sqrt{-x}$ and $y = \phi(x) = \sqrt{-x}$, $x \le 0$, corresponding to two branches of a parabola (see Fig. 1.1). Each branch plays the role of a sheet.

For a second example, let x be a two-dimensional vector $x = (x_1, x_2)$ and y a scalar.

In the case $g = y^2 + (x_1 + x_2)$, the equation

$$g = y^2 + (x_1 + x_2) = 0$$

gives two roots

$$y = \phi(x_1, x_2) = -\sqrt{-x_1 - x_2} \quad \text{and} \quad y = \phi(x_1, x_2) = \sqrt{-x_1 - x_2}, \quad x_1 + x_2 \le 0,$$

Fig. 1.1 Slow curve with two branches: $y \geq 0$, $y \leq 0$

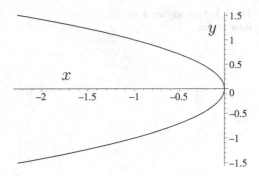

Fig. 1.2 Slow surface with two sheets: $y = \pm\sqrt{-x_1 - x_2}$, $x_1 + x_2 \leq 0$

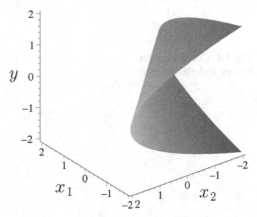

corresponding to two sheets of the slow surface (see Fig. 1.2).

As a third example, we take the case $g = y^3 - y(x_1^2 + x_2^2) - y$, and obtain three roots

$$y = \phi(x_1, x_2) = 0 \quad \text{and} \quad y = \phi(x_1, x_2) = \pm\sqrt{x_1^2 + x_2^2 + 1}$$

of the equation $g = y^3 - y(x_1^2 + x_2^2) - y = 0$ (see Fig. 1.3).

For the final example, we take $g = y^3 + x_1 y + x_2$ and obtain three sheets of the slow surface $y^3 + x_1 y + x_2 = 0$, corresponding to the three roots of this cubic equation in y (see Fig. 1.4).

We now return to integral manifolds of a differential equation. The motion along an integral manifold of the system (1.27) is governed by the equation

$$\dot{x} = f(x, h(x, t, \varepsilon), t, \varepsilon), \tag{1.31}$$

where $y = h(x, t, \varepsilon)$ is a slow integral manifold. If $x(t, \varepsilon)$ is a solution of (1.31), then the pair $\left(x(t, \varepsilon), y(t, \varepsilon)\right)$, where $y(t, \varepsilon) = h(x(t, \varepsilon), t, \varepsilon)$, is a solution of the original system (1.27), since it defines a trajectory on the integral manifold.

Fig. 1.3 Slow surface with
three sheets

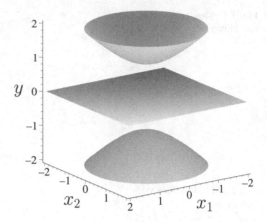

Fig. 1.4 Continuous slow
surface with three sheets

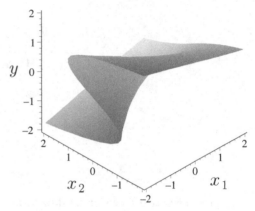

In an autonomous system

$$\frac{dx}{dt} = f(x, y, \varepsilon),$$

$$\varepsilon\frac{dy}{dt} = g(x, y, \varepsilon),$$
(1.32)

i.e., where f and g do not have an explicit dependence on t, an integral manifold
has the form $S_1 \times (-\infty, \infty)$, where S_1 is a surface in the phase space $\mathbb{R}^m \times \mathbb{R}^n$, i.e.,
the $x - y$ space, and the integral manifolds will be graphs of functions

$$y = h(x, \varepsilon).$$

In the case of an autonomous system the term "invariant manifold" is normally
used instead of "integral manifold". More exactly

Definition 3. A smooth surface S in $\mathbb{R}^m \times \mathbb{R}^n$ is called an invariant manifold of the system (1.32) if any trajectory of the system that has at least one point in common with S lies entirely on S.

Tikhonov has proved the following theorem [199], which answers the question about the permissibility of the application of the degenerate system (1.28) as a "zero approximation" to (1.27). The basic assumption of the theorem requires the asymptotic stability of the function $y = \phi(x,t)$ (recall that $y = \phi(x,t)$ is a solution of $g(x,y,t,0) = 0$) as a steady state solution of the so-called associated (or adjoined, or boundary layer) system

$$\frac{dy}{d\tau} = g(x,y,t,0) \qquad (1.33)$$

for all fixed x and t [formally, we set the variable $\tau = t/\varepsilon$, independent of the parameter t in (1.33)]: i.e., $y = \phi(x,t)$ is the asymptotically stable solution to (1.33) as $\tau \to \infty$. Equation (1.33) is sometimes called *the layer equation*.

We will formulate the statement of Tikhonov's theorem in a very simple form. We consider the system (1.27) with initial conditions $x(t_0, \varepsilon) = x_0$, $y(t_0, \varepsilon) = y_0$ for the scalar variable y. In order to quote a much-simplified version of Tikhonov's theorem the following conditions have to be satisfied:

(i) The functions f and g are uniformly continuous and bounded, together with their partial derivatives with respect to all variables in some open domain of space (x, y), $t \in [t_0, t_1]$ and $\varepsilon \in [0, \varepsilon_0]$.

(ii) The boundary layer equation (1.33) has a solution for a given initial value.

(iii) For every fixed x and t, $y = \phi(x,t)$ is an *isolated root* of $g(x,y,t,0) = 0$ i.e., $g(x, \phi(x,t), t, 0) = 0$, and there exists a positive number $\delta > 0$ such that the conditions $\|y - \phi(x,t)\| < \delta$ and $y \neq \phi(x,t)$ imply $g(x,y) \neq 0$.

   This does not mean that the equation $g(x,y,t,0) = 0$ has no other roots except $\phi(x,t)$.

(iv) The equation

$$\dot{x} = f(x, \phi(x,t), t, 0)$$

   with a given initial condition has a solution $x = \bar{x}(t)$ on $t \in [t_0, t_1]$.

(v) There exists $\gamma > 0$ such that $g_y(x, \phi(x,t), t, 0) \leq -\gamma$.

   This implies that $\phi(x,t)$ is an asymptotically stable equilibrium solution to (1.33).

(vi) The point y_0 belongs to the basin of attraction of the steady state solution $y = \phi(x_0, t_0)$.

Assumption (vi) identifies the initial points for which one can be sure that the solution to (1.27) converges to corresponding solution of (1.28). To understand its meaning, recall that the steady state of a nonlinear system does not necessarily attract all other solutions. If the steady state attracts all other solutions it is called *globally asymptotically stable*. Most often, only solutions originating from a neighborhood of a steady state converges to it as $t \to \infty$. Such a neighborhood is called a basin of attraction of steady state. Now, if we take an initial value x_0, then the steady state of the layer equation (1.33) is $\phi(x_0, t_0)$. To make sure that Tikhonov's theorem will work, we must take y_0 from the basin of attraction of $\phi(x_0, t_0)$.

It should be noted that the stability condition (v) is slightly stronger than the one given in [199]. Moreover, (v) guarantees that (vi) holds for all y_0 with sufficiently small $y_0 - \phi(x_0, t_0)$.

Then, the following result holds:

Theorem 1 (Tikhonov's Theorem). *If assumptions* (i)–(vi) *are valid then the solution* $(x(\tau, \varepsilon), y(\tau, \varepsilon))$ *of the initial value problem (1.27) exists in* $[t_0, t_1]$ *and the following conditions hold*

$$\lim_{\varepsilon \to 0} x(t, \varepsilon) = \bar{x}(t), \quad t_0 \le t \le t_1; \tag{1.34}$$

$$\lim_{\varepsilon \to 0} y(t, \varepsilon) = \phi(\bar{x}(t), t), \quad t_0 < t \le t_1. \tag{1.35}$$

The convergence in (1.34) and (1.35) is uniform in the interval $t_0 \le t \le t_1$ *for* $x(t, \varepsilon)$ *and in any interval* $t_0 < v \le t \le t_1$ *for* $y(t, \varepsilon)$.

This mean that, under the conditions of Tikhonov's theorem, the solution travels to the slow surface and is the limit of the exact solution as $\varepsilon \to 0$.

Tikhonov's theorem permits different interpretations. It is possible to consider his result as the first step in constructing the asymptotic expansion of the initial value problem. Now we can consider it as the first step in order reduction.

The foregoing is illustrated by the autonomous system (1.22). Setting $\varepsilon = 0$, we obtain *the degenerate problem*:

$$\begin{aligned} \dot{x} &= y, \ x(0) = x_0, \\ 0 &= -ax - by, \end{aligned} \tag{1.36}$$

which cannot satisfy the initial condition $y(0) = y_0$.

In this case the role of the slow surface is played by the *slow curve* which is described by the equation

$$0 = -ax - by.$$

In the case $b > 0$ the root $y = -ax/b = \lambda_1(0)x$ of this equation is the asymptotically stable steady state solution as $\tau \to \infty$ to the corresponding boundary layer equation

$$\frac{dy}{d\tau} = -ax - by.$$

Even though the initial condition $y(0) = y_0$ is not satisfied, the approximation $y = -ax/b$ is yet the stable steady state solution of the second equation in (1.22) as $t \to \infty$. This follows from the fact that y_0 belongs to the basin of attraction of the steady state solution. Here $\lambda_1(0) = \lambda_1(\varepsilon)\big|_{\varepsilon=0}$ is given by (1.5) and (1.7). This means that Tikhonov's theorem is applicable to the system (1.22), and the solution of (1.22) tends to the solution of (1.36) as $\varepsilon \to 0$. Of course, the identical conclusion can be derived from the exact solution of (1.22) under $\varepsilon \to 0$.

There are many applied problems where the use of the degenerate equations, obtained by setting $\varepsilon = 0$, instead of the full equations give acceptable results, but in some cases the approximation (1.28) is too crude. Readers who have an interest in such problems are referred to Chap. 7 in [117].

There are at least two means of proceeding from Eq. (1.28) as an approximation to (1.27). In the first the validity of proceeding to the limit

$$x(t,\varepsilon) \to x_0(t), \quad y(t,\varepsilon) \to y_0(t) \text{ as } \varepsilon \to 0$$

is examined, where $x = x(t,\varepsilon)$, $y = y(t,\varepsilon)$ are solutions to the Eq. (1.27), and $x_0(t)$, $y_0(t)$ are solutions to the degenerate problem, i.e. $x_0(t)$ is a solution of the equation

$$\frac{dx}{dt} = f(x,\phi(x,t),t,0), \tag{1.37}$$

and $y_0(t) = \phi(x_0(t),t)$ is a solution of $0 = g(x_0(t),y,t,0)$. If the approximation $x = x_0(t)$, $y = y_0(t)$ is too crude, it is reasonable to construct more exact approximations for the functions $x(t,\varepsilon)$, $y(t,\varepsilon)$ with the help of asymptotic methods, e.g. the boundary layer method [127,130,204,205,211], the multiple-scale method [121], the regularization method [99], the averaging method [14].

The second method considers the degenerate equation (1.28) as the zero approximation of the decomposition of the system (1.27), where the slow variable x is constructed from the independent equation (1.37), and the fast variable y is determined either from the algebraic relation $y = \phi(x,t)$, or from the associated (or boundary layer) equation (1.33). From this point of view the more exact the decomposition of the system is, the more precise is the result. This means that the independent equation (1.31) for the slow variable x is designed to have greater accuracy, and the fast variable is determined from a more precise algebraic relation of the form $y = h(x,t,\varepsilon)$, or from some differential equation of the dimension $n = \dim y$, whose coefficients may depend on the slow variable. This second means of proceeding is the basis for the approach developed in this book.

1.3.2 Integral Manifolds and Asymptotic Expansions of Solutions

Returning to the system (1.22), we see that an important role is played by two trajectories, viz., the straight lines $y = \lambda_1(\varepsilon)x$ and $y = \lambda_2(\varepsilon)x$. The line $y = \lambda_1(\varepsilon)x$ can be considered to be a *slow integral manifold*, because if we choose an initial point (x_0, y_0) on this line, i.e. $y_0 = \lambda_1 x_0$ or $C_2 = 0$ in (1.10), than the whole trajectory of the corresponding solution

$$x = \frac{\lambda_2 x_0 - y_0}{\lambda_2 - \lambda_1} e^{\lambda_1 t} = x_0 e^{\lambda_1 t}, \quad y = \dot{x} = \lambda_1 \frac{\lambda_2 x_0 - y_0}{\lambda_2 - \lambda_1} e^{\lambda_1 t} = \lambda_1 x$$

lies on this straight line. The behavior of solutions $x = x(t, \varepsilon)$ on the slow integral manifold is then described by the first order differential equation

$$\dot{x} = \lambda_1 x.$$

Using the asymptotic representations

$$\lambda_1 = -a/b - \varepsilon a^2/b^3 + O(\varepsilon^2),$$
$$\lambda_2 = \varepsilon^{-1}[-b + \varepsilon a/b + O(\varepsilon^2)]$$

it is easy to see that the invariant line $y = \lambda_1(\varepsilon)x$ is *attractive* when $b > 0$ and *repulsive* when $b < 0$ for any a. This is also readily seen from the exact solution given above or from (1.10) and (1.11). Note that the solution of the degenerate problem (1.36) can be considered to be *a limiting solution* (as $\varepsilon \to 0$) with respect to the solution of the original problem (1.22) when $b > 0$, see Tikhonov's theorem above.

Any trajectory of (1.22) can be represented as a trajectory on the attractive slow integral manifold plus *an asymptotically negligible* term corresponding to $\lambda_2(\varepsilon)$ when $b > 0$ (see Fig. 1.5 which demonstrates that trajectories go through the slow curve and approach the slow integral manifold).

In a similar manner, we to say that line $y = \lambda_2(\varepsilon)x$ is *the fast invariant manifold* [170]. The trajectory $y = \lambda_2(\varepsilon)x$ corresponds to the initial condition $y_0 = \lambda_2(\varepsilon)x_0$, or $C_1 = 0$ in (1.10).

As a note of warning, the formal use of an asymptotic expansion can lead to an incorrect representation of the solution, or to a representation with restricted

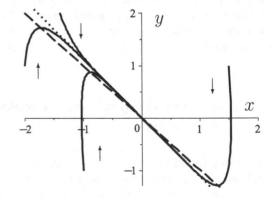

Fig. 1.5 Four trajectories, corresponding to different initial conditions, (*solid lines*) going through the slow curve $y = \lambda_1(0)x$ (*dashed line*) and approach the slow invariant manifold $y = \lambda_1(\varepsilon)x$ (*dotted straight line*) ($a = b = 1$, $\varepsilon = 0.1$), where the *arrows* indicate increasing time

application. By way of example, consider the second order linear differential equation with constant coefficients

$$\varepsilon\ddot{x} + (1 + \varepsilon^2)\dot{x} + \varepsilon x = 0; \quad x(0) = x_0, \quad \dot{x}(0) = y_0.$$

It is a straightforward exercise to check that the exact solution of this equation is

$$x = x_{outer}(t, \varepsilon) + x_{layer}(\tau, \varepsilon),$$

where

$$x_{outer}(t, \varepsilon) = \frac{(\varepsilon y_0 + x_0)}{1 - \varepsilon^2} e^{-\varepsilon t},$$

$$x_{layer}(\tau, \varepsilon) = -\frac{\varepsilon(y_0 + \varepsilon x_0)}{1 - \varepsilon^2} e^{-t/\varepsilon}.$$

The outer solution is an exponentially decreasing function, whereas the asymptotic expansion of this solution in terms of powers of ε

$$x_{outer}(t, \varepsilon) = x_0 + \varepsilon(y_0 - x_0 t) + \varepsilon^2(x_0 t^2/2 - y_0 t + x_0) + O(\varepsilon^3 t^3)$$

has a polynomial behaviour, and any order asymptotic approximation of the outer solution tends to infinity (plus or minus) as $t \to +\infty$. The powers of t in the expansion of $x_{outer}(t, \varepsilon)$ are called "secular" terms and the expansion is not uniformly valid in t. This indicates a different form of asymptotic expansion is required, e.g., multiple scales. Such difficulties emerge because asymptotic expansions of this kind are adequate only for a finite range of t, i.e., $t = o(\varepsilon^{-1})$. A multiple scale approach would normally avoid this difficulty, as is shown shortly.

If we consider this second order equation in the form of a planar differential system

$$\dot{x} = y,$$

$$\varepsilon\dot{y} = -(1 + \varepsilon^2)y - \varepsilon x,$$

then the solution is

$$x = x_{outer} + x_{layer} = \frac{(\varepsilon y_0 + x_0)}{1 - \varepsilon^2} e^{-\varepsilon t} - \frac{\varepsilon(y_0 + \varepsilon x_0)}{1 - \varepsilon^2} e^{-t/\varepsilon},$$

$$y = y_{outer} + y_{layer} = -\varepsilon\frac{(\varepsilon y_0 + x_0)}{1 - \varepsilon^2} e^{-\varepsilon t} + \frac{y_0 + \varepsilon x_0}{1 - \varepsilon^2} e^{-t/\varepsilon}.$$

It is clear from x_{outer} and y_{outer} that the straight line

$$y = -\varepsilon x$$

plays the role of an attractive slow invariant manifold, the motion on which is described by the equation

$$\dot{x} = -\varepsilon x.$$

The solution of this equation, with the initial condition $x = \frac{(\varepsilon y_0 + x_0)}{1-\varepsilon^2}$ as $t = 0$, is precisely

$$x = x_{\text{outer}} = \frac{(\varepsilon y_0 + x_0)}{1 - \varepsilon^2} e^{-\varepsilon t}.$$

In the context of this approach, based on the integral manifold method, the essence is to study the behaviour of solutions on the slow integral manifold.

Note that the method of multiple scales is able to give an acceptable result in this case. Introduce $\tau = t/\varepsilon$, then the equation becomes

$$\frac{d^2 x}{d\tau^2} + (1 + \varepsilon^2)\frac{dx}{d\tau} + \varepsilon^2 ax = 0.$$

Let $x(\tau, \varepsilon) = F(\tau, s, \varepsilon)$, where $s = \varepsilon t = \varepsilon^2 \tau$, and assume the expansion

$$F(\tau, t; \varepsilon) = F_0(\tau, s) + \varepsilon^2 F_1(\tau, s) + \varepsilon^4 F_2(\tau, s) + \dots .$$

On substituting this expansion into the last differential equation and equating powers of ε we get the sequence of problems.

ε^0 :

$$\frac{\partial^2 F_0}{\partial \tau^2} + \frac{\partial F_0}{\partial \tau} = 0, \quad F_0(0, 0) = x_0, \quad \frac{\partial F_0}{\partial \tau}(0, 0) = 0.$$

Thus

$$F_0(\tau, s) = A_0(s)e^{-\tau} + B_0(s),$$

where the initial conditions imply

$$A_0(0) = 0, \quad A_0(0) + B_0(0) = x_0.$$

ε^2 :

$$\frac{\partial^2 F_1}{\partial \tau^2} + \frac{\partial F_1}{\partial \tau} = -F_0 - 2\frac{\partial^2 F_0}{\partial \tau \partial s} - \frac{\partial F_0}{\partial \tau} - \frac{\partial F_0}{\partial s}.$$

Then the r.h.s. is $A_0'(s)e^{-\tau} - (B_0'(s) + B_0(s))$.

To avoid secular terms in τ, require

$$A_0'(s) = 0, \quad \text{and } B_0'(s) + B_0(s) = 0.$$

So $A_0(s) \equiv 0$ and $B_0(s) = x_0 e^{-s}$ on using the initial conditions.
Then

$$x(t, \varepsilon) = F_0(\tau, s) + O(\varepsilon) = x_0 e^{-\varepsilon t} + O(\varepsilon),$$

i.e. the slow component of the solution is given by $y = \dot{x} = -\varepsilon x$, as before.

The application of integral manifolds in the investigation of singularly perturbed systems aims to replace the analysis of the full system by the analysis of a system of lower dimension. The lowering of the dimension occurs due to the decomposition of the initial system in the vicinity of the integral surface into the independent slow subsystem of the form (1.31) and the fast subsystem. If the slow integral manifold is attracting, then the analysis of the system under consideration can be replaced by the analysis of the slow subsystem.

Note in conclusion that a slow integral manifold is associated with outer (slow) solutions, and a fast integral manifold is associated with boundary layer (fast) corrections.

Chapter 2
Slow Integral Manifolds

Abstract In the present chapter we use a method for the qualitative asymptotic analysis of singularly perturbed differential equations by reducing the order of the differential system under consideration. The method relies on the theory of integral manifolds. It essentially replaces the original system by another system on an integral manifold with a lower dimension that is equal to that of the slow subsystem. The emphasis in this chapter is on the study of autonomous systems.

2.1 Introduction

The non-autonomous system is

$$\frac{dx}{dt} = f(x, y, t, \varepsilon),$$

$$\varepsilon \frac{dy}{dt} = g(x, y, t, \varepsilon),$$

(2.1)

and the autonomous system is

$$\dot{x} = f(x, y, \varepsilon),$$

$$\varepsilon \dot{y} = g(x, y, \varepsilon).$$

(2.2)

Definition 4. A smooth surface S in $\mathbb{R}^m \times \mathbb{R}^n$ is called an invariant manifold of the system (2.2) if any trajectory of the system that has at least one point in common with S lies entirely on S.

This means that any trajectory $x = x(t, \varepsilon)$, $y = y(t, \varepsilon)$ of the system (2.2) that has at least one point $x = x_0$, $y = y_0$ in common with the invariant surface $y = h(x, \varepsilon)$, i.e. $y_0 = h(x_0, \varepsilon)$, then it lies entirely on the invariant surface, i.e. on $y(t, \varepsilon) = h(x(t, \varepsilon), \varepsilon)$.

The motion along an invariant manifold of the autonomous system (2.2) is governed by the equation

$$\dot{x} = f(x, h(x, \varepsilon), \varepsilon).$$

(2.3)

© Springer International Publishing Switzerland 2014

E. Shchepakina et al., *Singular Perturbations*, Lecture Notes in Mathematics 2114,
DOI 10.1007/978-3-319-09570-7_2

If $x(t, \varepsilon)$ is a solution of this equation, then the pair $\left(x(t, \varepsilon), y(t, \varepsilon)\right)$, where $y(t, \varepsilon) = h(x(t, \varepsilon), \varepsilon)$, is a solution of the original system (2.2), since it defines a trajectory on the invariant manifold.

Note that the formal substitution of the function $h(x, \varepsilon)$ instead y into the autonomous system (2.2) gives the first order PDE, the so called *invariance equation*,

$$\varepsilon \frac{\partial h}{\partial x} f(x, h(x, \varepsilon), \varepsilon) = g(x, h, \varepsilon) \tag{2.4}$$

for $h(x(t), \varepsilon)$, since $\varepsilon \dot{y} = \varepsilon \frac{\partial h}{\partial x} \dot{x}$.

In the case of a non-autonomous system, if any integral curve $\left(t, x(t, \varepsilon), y(t, \varepsilon)\right)$ of the solution $x = x(t, \varepsilon)$, $y = y(t, \varepsilon)$ to the system (2.1) has at least one point $x = x_0$, $y = y_0$ in common with the integral surface $y = h(x, t, \varepsilon)$, i.e. $y_0 = h(x_0, t_0, \varepsilon)$, then it lies entirely in this surface, i.e. $y(t, \varepsilon) = h(x(t, \varepsilon), t, \varepsilon)$.

The motion along an integral manifold of the non-autonomous system (2.1) is governed by the equation

$$\dot{x} = f(x, h(x, t, \varepsilon), t, \varepsilon).$$

In the non-autonomous case the invariance equation for $y = h(x, t, \varepsilon)$ is

$$\varepsilon \frac{\partial h}{\partial t} + \varepsilon \frac{\partial h}{\partial x} f(x, h(x, t, \varepsilon), t, \varepsilon) = g(x, h, \varepsilon). \tag{2.5}$$

Consider now the *the boundary layer* subsystem of (2.2), that is,

$$\frac{dy}{d\tau} = g(x, y, 0), \quad \tau = t/\varepsilon,$$

treating x as a vector parameter. We shall assume that some of the steady states $y^0 = y^0(x)$ of this subsystem, defined $g(x, y, 0) = 0$, are asymptotically stable and that a trajectory starting at any point of the basin of attraction approaches one of these states as closely as desired as $\tau \to \infty$. This assumption will hold, for example, if the matrix

$$B(x, t) \equiv (\partial g / \partial y)(x, y^0(x,), 0) \equiv \begin{pmatrix} \dfrac{\partial g_1}{\partial y_1} & \cdots & \dfrac{\partial g_1}{\partial y_n} \\ \cdots & \cdots & \cdots \\ \dfrac{\partial g_n}{\partial y_1} & \cdots & \dfrac{\partial g_n}{\partial y_n} \end{pmatrix} \Bigg|_{y = y^0(x)}$$

is stable for some of the stationary states and the basin of attraction can be represented as the union of the basins of attraction of the asymptotically stable steady states. We recall that a matrix is stable if its spectrum is located in the left open complex halfplane, i.e. all eigenvalues of this matrix have negative real parts.

Notwithstanding the fact that we are interested primarily in autonomous systems, all statements will be formulated in the more general case of non-autonomous systems.

It is assumed that

(I) The functions f, g and ϕ are uniformly continuous and bounded, together with their partial derivatives with respect to all variables up to the $(k + 2)$-order $(k \geq 0)$ for y in some open domain of space $\mathbb{R}^n$, $x \in \mathbb{R}^m$, $t \in [-\infty, \infty]$ and $\varepsilon \in [0, \varepsilon_0]$.

(II) The eigenvalues $\lambda_i(x, t)(i = 1, \ldots, n)$ of the matrix $B(x, t) = g_y(x, \phi(x, t), t, 0)$ satisfy the inequality

$$Re\lambda_i(x, t) \leq -2\gamma < 0, \tag{2.6}$$

for some $\gamma > 0$.

Recall $\phi(x, t)$ is a root of the equation $g(x, \phi(x, t), t, 0) = 0$.

Then the following result holds (see e.g. [92, 170, 195]):

Proposition 1. *Under the assumptions* (I) *and* (II) *there is a sufficiently small positive* ε_1, $\varepsilon_1 \leq \varepsilon_0$, *such that, for* $\varepsilon \in I_1 := \{\varepsilon \in \mathbb{R} : 0 < \varepsilon < \varepsilon_1\}$, *the system* (2.1) *has a smooth integral manifold* $\mathcal{M}_\varepsilon$ *with the representation*

$$\mathcal{M}_\varepsilon := \{(x, y, t) \in \mathbb{R}^{m+n+1} : y = h(x, t, \varepsilon), (x, t) \in G \times \mathbb{R}\},$$

for some domain $G \in \mathbb{R}^m$.

Proposition 1 guarantees that the invariance equation (2.5) can yield $y = h(x, t, \varepsilon)$ which is the slow integral manifold.

Remark 2.1. The global boundedness assumption in (I) with respect to (x, y) can be relaxed by modifying f and g outside some bounded region of $\mathbb{R}^n \times \mathbb{R}^m$.

We will present the proof of this Proposition in the Appendix in Chap. 9 (see Theorem 4).

2.2 Stability of Slow Integral Manifolds

In applications it is often assumed that the spectrum of the Jacobian matrix

$$g_y(x, \phi(x, t), t, 0)$$

is located in the left half plane, where $\phi(x, t)$ is the root of the equation $g(x, \phi, t, 0) = 0$. Under this additional hypothesis the manifold $\mathcal{M}_\varepsilon$ is exponentially attracting for $\varepsilon \in I_1$. This means: the solution $x = x(t,\varepsilon)$, $y = y(t,\varepsilon)$ of the original system (2.1) that satisfied the initial condition $x(t_0,\varepsilon) = x^0$, $y(t_0,\varepsilon) = y^0$ can be represented as

$$
\begin{aligned}
x(t,\varepsilon) &= v(t,\varepsilon) + \varepsilon\varphi_1(t,\varepsilon), \\
y(t,\varepsilon) &= \bar{y}(t,\varepsilon) + \varphi_2(t,\varepsilon).
\end{aligned}
\tag{2.7}
$$

The fact that there is ε before φ_1 in the first equation and no ε before φ_2 in the second one is in agreement with the statement of Tikhonov's theorem (see Sect. 1.3.1). It is possible to prove that there exists a point v^0 which is the initial value for the motion along an integral manifold which is a solution $v(t,\varepsilon)$ of the equation $\dot{v} = f(v,h(v,t,\varepsilon),t,\varepsilon)$. The functions $\varphi_1(t,\varepsilon)$, $\varphi_2(t,\varepsilon)$ are corrections that determine the degree to which trajectories passing near the manifold $\mathcal{M}_\varepsilon$ tend asymptotically to the corresponding trajectories on the manifold as t increases. They satisfy the following inequalities:

$$
|\varphi_i(t,\varepsilon)| \le N|y^0 - h(x^0,t_0,\varepsilon)|\exp[-\gamma(t-t_0)/\varepsilon], \quad i = 1,2,
\tag{2.8}
$$

and $|y^0 - h(x^0,t_0,\varepsilon)| \le \rho$ for some positive ρ and $t \ge t_0$. An application of this result to a problem on high-gain control is given in Sect. 5.5. As an illustration of the above we consider the following example.

Example 1. Consider the third order linear differential equation

$$
\varepsilon\frac{d^3x}{dt^3} + \frac{d^2x}{dt^2} - 2\frac{dx}{dt} + 4(1 + 2\varepsilon)x = 0,
\tag{2.9}
$$

with constant coefficients and initial conditions

$$
x(0) = x_0, \quad \dot{x}(0) = \dot{x}_0, \quad \ddot{x}(0) = \ddot{x}_0.
\tag{2.10}
$$

This equation is rewritten in the form of the differential system

$$
\dot{x}_1 = x_2,
\tag{2.11}
$$

$$
\dot{x}_2 = y,
\tag{2.12}
$$

$$
\varepsilon\dot{y} = -y + 2x_2 - 4(1 + 2\varepsilon)x_1,
\tag{2.13}
$$

where $x_1 = x$, with initial conditions

$$
x_1(0) = x_0, \quad x_2(0) = \dot{x}_0, \quad y(0) = \ddot{x}_0.
\tag{2.14}
$$

The slow surface for this system ($\varepsilon = 0$) takes the form

$$
y = -4x_1 + 2x_2,
$$

and the slow invariant manifold may be written in the same form. To check this fact it is necessary to write down the invariance equation (2.4) for $y = h(x_1, x_2) = -4x_1 + 2x_2$:

$$\varepsilon[-4x_2 + 2(-4x_1 + 2x_2)] = 4x_1 - 2x_2 + 2x_2 - 4(1 + 2\varepsilon)x_1,$$

which is an identity. Here

$$x = \begin{pmatrix} x_1 \\ x_2 \end{pmatrix}, \quad f = \begin{pmatrix} x_2 \\ y \end{pmatrix},$$

$g = -y + 2x_2 - 4(1 + 2\varepsilon)x_1$ and $\frac{\partial h}{\partial x} = (-4, 2)$.

Introducing the new variable z by the formula

$$z = y + 4x_1 - 2x_2, \tag{2.15}$$

we obtain the initial value problem for z:

$$\varepsilon \dot{z} = -(1 + 2\varepsilon)z, \quad z(0) = z_0 = y(0) + 4x_1(0) - 2x_2(0) = \ddot{x}_0 + 4x_0 - 2\dot{x}_0,$$

the solution to which is

$$z = z(t, \varepsilon) = z_0 \exp\left(-\frac{1 + 2\varepsilon}{\varepsilon}t\right). \tag{2.16}$$

Now we obtain the following differential system for x_1 and x_2:

$$\dot{x}_1 = x_2, \quad \dot{x}_2 = -4x_1 + 2x_2 + z(t, \varepsilon).$$

It is a straightforward exercise to check that

$$
\begin{aligned}
x_1 = x_1(t, \varepsilon) = & \left[\left(x_0 - \frac{\varepsilon^2}{\zeta}z_0\right)\cos(\sqrt{3}t)\right. \\
& \left. + \frac{1}{\sqrt{3}}\left(\dot{x}_0 - x_0 + \frac{\varepsilon(1 + 3\varepsilon)}{\zeta}z_0\right)\sin(\sqrt{3}t)\right]e^t \\
& + \frac{\varepsilon^2}{\zeta}z_0 \exp\left(-\frac{1 + 2\varepsilon}{\varepsilon}t\right), \\
x_2 = x_2(t, \varepsilon) = & \left[\left(\dot{x}_0 + \frac{\varepsilon(1 + 2\varepsilon)}{\zeta}z_0\right)\cos(\sqrt{3}t)\right.
\end{aligned}
$$

$$+\frac{1}{\sqrt{3}}\left(\dot{x}_0 - 4x_0 + \frac{\varepsilon(1+6\varepsilon)}{\zeta}z_0\right)\sin(\sqrt{3}t)\right]e^t$$

$$-\frac{\varepsilon(1+2\varepsilon)}{\zeta}z_0\exp\left(-\frac{1+2\varepsilon}{\varepsilon}t\right),$$

where

$$\zeta = \zeta(\varepsilon) = 1 + 6\varepsilon + 12\varepsilon^2.$$

It follows from (2.15) that

$$y = y(t,\varepsilon) = 2\left[\left(\dot{x}_0 - 2x_0 + \frac{\varepsilon(1+4\varepsilon)}{\zeta}z_0\right)\cos(\sqrt{3}t)\right.$$

$$\left.-\frac{1}{\sqrt{3}}\left(\dot{x}_0 + 2x_0 + \frac{\varepsilon}{\zeta}z_0\right)\sin(\sqrt{3}t)\right]e^t$$

$$+\frac{(1+2\varepsilon)^2}{\zeta}z_0\exp\left(-\frac{1+2\varepsilon}{\varepsilon}t\right).$$

Thus, we obtain the representation (2.7) for (2.11)–(2.13) in the form

$$x_1(t,\varepsilon) = v_1(t,\varepsilon) + \varepsilon^2\varphi_{11}(t,\varepsilon),$$

$$x_2(t,\varepsilon) = v_2(t,\varepsilon) + \varepsilon\varphi_{12}(t,\varepsilon),$$

$$y(t,\varepsilon) = \bar{y}(t,\varepsilon) + \varphi_2(t,\varepsilon).$$

Here

$$v_1(t,\varepsilon) = \left[\left(x_0 - \frac{\varepsilon^2}{\zeta}z_0\right)\cos(\sqrt{3}t) + \frac{1}{\sqrt{3}}\left(\dot{x}_0 - x_0 + \frac{\varepsilon(1+3\varepsilon)}{\zeta}z_0\right)\sin(\sqrt{3}t)\right]e^t,$$

$$v_2(t,\varepsilon) = \left[\left(\dot{x}_0 + \frac{\varepsilon(1+2\varepsilon)}{\zeta}z_0\right)\cos(\sqrt{3}t)\right.$$

$$\left.+\frac{1}{\sqrt{3}}\left(\dot{x}_0 - 4x_0 + \frac{\varepsilon(1+6\varepsilon)}{\zeta}z_0\right)\sin(\sqrt{3}t)\right]e^t,$$

$$\bar{y}(t,\varepsilon) = 2\left[\left(\dot{x}_0 - 2x_0 + \frac{\varepsilon(1+4\varepsilon)}{\zeta}z_0\right)\cos(\sqrt{3}t)\right.$$

$$\left.-\frac{1}{\sqrt{3}}\left(\dot{x}_0 + 2x_0 + \frac{\varepsilon}{\zeta}z_0\right)\sin(\sqrt{3}t)\right]e^t,$$

and

$$\varphi_{11}(t,\varepsilon) = \frac{1}{\zeta} z_0 \exp\left(-\frac{1+2\varepsilon}{\varepsilon}t\right),$$

$$\varphi_{12}(t,\varepsilon) = -\frac{(1+2\varepsilon)}{\zeta} z_0 \exp\left(-\frac{1+2\varepsilon}{\varepsilon}t\right),$$

$$\varphi_2(t,\varepsilon) = \frac{(1+2\varepsilon)^2}{\zeta} z_0 \exp\left(-\frac{1+2\varepsilon}{\varepsilon}t\right).$$

Note that the solutions

$$x_1 = x_1(t,\varepsilon), \quad x_2 = x_2(t,\varepsilon), \quad y = y(t,\varepsilon),$$

which satisfy the initial conditions

$$x_1(0,\varepsilon) = x_0, \quad x_2(0,\varepsilon) = \dot{x}_0, \quad y(0,\varepsilon) = y_0,$$

are exponentially attracted to the corresponding solutions

$$x_1 = v_1(t,\varepsilon), \quad x_2 = v_2(t,\varepsilon), \quad y = \bar{y}(t,\varepsilon)$$

on the slow invariant manifolds as $t \to \infty$ (see Fig. 2.1). Further, note that the initial conditions for v_1 and v_2 are

$$v_1(0,\varepsilon) = v_1^0 = x_0 - \frac{\varepsilon^2}{\zeta} z_0, \quad v_2(0,\varepsilon) = v_2^0 = \dot{x}_0 + \frac{\varepsilon(1+2\varepsilon)}{\zeta} z_0.$$

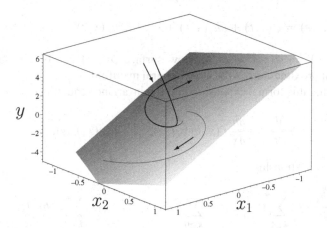

Fig. 2.1 Example 1: the slow invariant manifold (*shaded plane*), two trajectories (the *thin solid lines*) on the slow invariant manifold and one trajectory outside it (the *thick solid line*) approaching the trajectory on slow invariant manifold with corresponding initial point; $\varepsilon = 0.1$. The *arrows* indicate increasing time

Then there are the initial values v_1^0 and v_2^0, rather than x_0 and $\dot{x}_0$, that provide the initial state for the solutions $v_1(t, \varepsilon)$ and $v_2(t, \varepsilon)$ on the slow invariant manifold.

From (2.7) and (2.8) we obtain the following *Lyapunov Reduction Principle* for a stable integral manifold defined by a function $y = h(x, t, \varepsilon)$. A solution $x = x(t, \varepsilon)$, $y = h(x(t, \varepsilon), t, \varepsilon)$ of the original non-autonomous system (2.1) is stable (asymptotically stable, unstable) if and only if the corresponding solution of the system of equations

$$\dot{v} = F(v, t, \varepsilon) = f(v, h(v, t, \varepsilon), t, \varepsilon)$$

on the integral manifold is stable (asymptotically stable, unstable) [170]. The Lyapunov Reduction Principle was extended to ordinary differential systems with Lipschitz right-hand sides by Pliss [138], and to singularly perturbed systems in [170, 197]. Thanks to the reduction principle and the representation (2.7), the qualitative behavior of trajectories of the original system near the integral manifold may be investigated by analyzing the equations on the manifold.

The Reduction Principle and the representation (2.7) can be found in [170].

2.3 Asymptotic Representation of Integral Manifolds

When the method of integral manifolds is being used to solve a specific problem, then a central question is the calculation of the function $h(x, t, \varepsilon)$ in terms of the manifold described. An exact calculation is generally impossible, and various approximations are necessary. One possibility is the asymptotic expansion of $h(x, t, \varepsilon)$ in integer powers of the small parameter ε:

$$h(x, t, \varepsilon) = \phi(x, t) + \varepsilon h_1(x, t) + \cdots + \varepsilon^k h_k(x, t) + \dots . \tag{2.17}$$

Here $h(x, t, 0) = \phi(x, t)$, i.e. the slow surface $\phi(x, t)$ can be considered as a zero-order approximations of the slow integral manifold.

Substituting this formal expansion into the invariance equation (2.5) i.e.,

$$\varepsilon \frac{\partial h}{\partial t} + \varepsilon \frac{\partial h}{\partial x} f(x, h(x, t, \varepsilon), t, \varepsilon) = g(x, h, \varepsilon), \tag{2.18}$$

we obtain the relationship

$$\varepsilon \sum_{k \geq 0} \varepsilon^k \frac{\partial h_k}{\partial t} + \varepsilon \sum_{k \geq 0} \varepsilon^k \frac{\partial h_k}{\partial x} f\left(x, \sum_{k \geq 0} \varepsilon^k h_k, t, \varepsilon\right) = g\left(x, \sum_{k \geq 0} \varepsilon^k h_k, t, \varepsilon\right). \tag{2.19}$$

We use the formal asymptotic representations

$$f\left(x, \sum_{k \geq 0} \varepsilon^k h_k, t, \varepsilon\right) = \sum_{k \geq 0} \varepsilon^k f_k(x, \phi, h_1, \ldots, h_k, t)$$

$$= f_0(x, \phi, t) + \varepsilon f_1(x, \phi, h_1, t) + \cdots$$

$$+ \varepsilon^k f_k(x, \phi, \ldots, h_k, t) + \cdots, \tag{2.20}$$

and

$$g\left(x, \sum_{k \geq 0} \varepsilon^k h_k, t, \varepsilon\right) = B(x, t) \sum_{k \geq 1} \varepsilon^k h_k + \sum_{k \geq 1} \varepsilon^k g_k(x, \phi, h_1, \ldots, h_{k-1}, t)$$

$$= B(x, t)(\varepsilon h_1 + \varepsilon^2 h_2 + \cdots + \varepsilon^k h_k + \cdots) \tag{2.21}$$

$$+ \varepsilon g_1(x, \phi, t) + \varepsilon^2 g_2(x, \phi, h_1, t)$$

$$+ \cdots + \varepsilon^k g_k(x, \phi, \ldots, h_{k-1}, t) + \cdots,$$

where the matrix $B(x, t) \equiv (\partial g / \partial y)(x, \phi, t, 0)$, and where

$$g(x, \phi(x, t), t, 0) = 0.$$

Substituting these formal expansions into (2.19)

$$\varepsilon \frac{\partial \phi}{\partial t} + \varepsilon^2 \frac{\partial h_1}{\partial t} + \cdots + \varepsilon^k \frac{\partial h_{k-1}}{\partial t}$$

$$+ \cdots + \left(\varepsilon \frac{\partial \phi}{\partial x} + \varepsilon^2 \frac{\partial h_1}{\partial x} + \cdots + \varepsilon^k \frac{\partial h_{k-1}}{\partial x} + \cdots\right)(f_0(x, \phi, t) + \varepsilon f_1(x, \phi, h_1, t)$$

$$+ \cdots + \varepsilon^k f_k(x, \phi, \ldots, h_k, t) + \cdots) = B(x, t)(\varepsilon h_1 + \varepsilon^2 h_2 + \cdots + \varepsilon^k h_k + \cdots)$$

$$+ \varepsilon g_1(x, \phi, t) + \varepsilon^2 g_2(x, \phi, h_1, t) + \cdots + \varepsilon^k g_k(x, \phi, \ldots, h_{k-1}, t) + \cdots,$$

and equating powers of ε, we obtain

$$\frac{\partial \phi}{\partial t} + \frac{\partial \phi}{\partial x} f_0(x, \phi, t) = Bh_1 + g_1,$$

$$\frac{\partial h_1}{\partial t} + \frac{\partial \phi}{\partial x} f_1 + \frac{\partial h_1}{\partial x} f_0 = Bh_2 + g_2,$$

$$\cdots$$

$$\frac{\partial h_{k-1}}{\partial t} + \sum_{0 \leq p \leq k-1} \frac{\partial h_p}{\partial x} f_{k-1-p} = Bh_k + g_k, \quad k = 2, 3, \ldots.$$

By virtue of (2.6), B is invertible and then

$$h_1 = B^{-1} \left[\frac{\partial \phi}{\partial t} + \frac{\partial \phi}{\partial x} f_0(x, \phi, t) - g_1 \right],$$ (2.22)

$$h_2 = B^{-1} \left[\frac{\partial h_1}{\partial t} + \frac{\partial \phi}{\partial x} f_1 + \frac{\partial h_1}{\partial x} f_0 - g_2 \right].$$ (2.23)

In general

$$h_k = B^{-1} \left[\frac{\partial h_{k-1}}{\partial t} + \sum_{0 \le p \le k-1} \frac{\partial h_p}{\partial x} f_{k-1-p} - g_k \right], k = 2, 3, \ldots$$ (2.24)

We recall that $\phi(x, t)$ is determined by $g(x, \phi(x, t), t, 0) = 0$. Now we can calculate an approximation to $h(x, t, \varepsilon)$ from (2.17).

Asymptotic expansions of slow integral manifolds were used in [186, 195, 196]. These papers address questions in gyroscopic systems, rotating bodies and orientation of satellites.

The justification of the asymptotic formulae will be given in the Appendix, see Chap. 9.

2.4 Two Mathematical Examples

We give two examples, for which exact solutions may be constructed, to illustrate slow invariant manifolds.

Example 2. The autonomous system of the two nonlinear scalar equations

$$\dot{x} = x, \quad \varepsilon \dot{y} = -y - x^2,$$

with the initial value conditions

$$x(0) = x_0, \quad y(0) = y_0,$$

has the exact solution

$$x(t, \varepsilon) = x_0 e^t, \quad y(t, \varepsilon) = -\frac{x_0^2}{1 + 2\varepsilon} e^{2t} + (y_0 + \frac{x_0^2}{1 + 2\varepsilon}) e^{-t/\varepsilon}.$$

The first term in $y(t, \varepsilon)$ is the outer solution, and the next term is the inner, or boundary layer, the part of solution.

This system possesses the attractive slow invariant manifold (see Definitions 1 and 3 in Sect. 1.3.1)

$$y = -\frac{x^2}{1 + 2\varepsilon},$$

because the trajectory on this manifold can be represented in the form

$$x(t, \varepsilon) = x_0 e^t, \quad y(t, \varepsilon) = -x(t, \varepsilon)^2/(1 + 2\varepsilon),$$

if we neglect terms of order $O(e^{-t/\varepsilon})$, i.e. the boundary layer terms are neglected.

If we use the formal procedure described above we have $f = x$, $g = -y - x^2$. Then the equation for the slow curve $\phi(x, t)$ (see Definition 2 in Sect. 1.3.1) is

$$0 = -y - x^2,$$

which has unique root $y = -x^2$. This root is stable because

$$\frac{\partial}{\partial y}(-y - x^2)\bigg|_{y=-x^2} = -1 < 0.$$

The invariance equation (2.4) for $h(x, \varepsilon)$ is

$$\varepsilon \frac{\partial h}{\partial x} x = -h - x^2,$$

and it is a straightforward exercise to check that the asymptotic expansion

$$h(x, \varepsilon) = \phi(x) + \varepsilon h_1(x) + \cdots + \varepsilon^k h_k(x) + \ldots$$

yields $\phi(x) = -x^2$, $h_1(x) = 2x^2$, $h_2(x) = -4x^2$, etc and coincides with the corresponding asymptotic expansion for the function $-x^2/(1 + 2\varepsilon)$.

The exact slow invariant manifold $h = -x^2/(1 + 2\varepsilon)$ and its zero order, $\phi(x) = -x^2$, and first order, $\phi(x) + \varepsilon h_1(x) = -x^2 + \varepsilon 2x^2$, approximations are shown in Figs. 2.2, 2.3 and 2.4. This shows how the approximations improve as $\varepsilon \to 0$.

Example 3. We consider the autonomous second order, nonlinear system

$$\dot{x} = y, \quad \varepsilon \dot{y} = -y - y^2,$$

with the initial conditions

$$x(0) = x_0, \quad y(0) = y_0.$$

Since the y equation may be written as

$$\left(\frac{1}{y} + \frac{1}{1 + y}\right) dy = -\frac{dt}{\varepsilon},$$

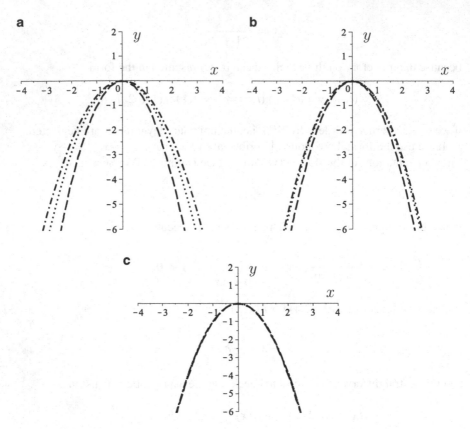

Fig. 2.2 Example 2: the exact slow invariant manifold h (the *dotted line*), the slow curve ϕ (the *dashed line*) and the first-order approximation to the slow invariant manifold (the *dashed-dotted line*); (**a**) $\varepsilon = 0.2$, (**b**) $\varepsilon = 0.1$, (**c**) $\varepsilon = 0.01$

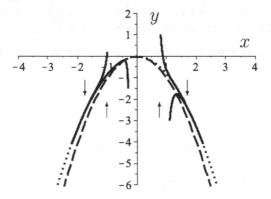

Fig. 2.3 Example 2: the slow invariant manifold (the *dotted line*), the slow curve (the *dashed line*) and the trajectories (the *solid lines*) with various initial points and $\varepsilon = 0.1$. The *arrows* indicate increasing time. The reader should note that even though the initial point is not on the slow invariant manifold, the solution eventually lies very close to the manifold

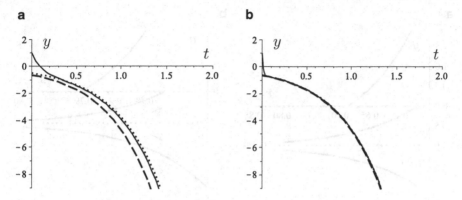

Fig. 2.4 Example 2: the y-component of the solution with $x_0 = 0.8$, $y_0 = 1.0$ (the *solid line*), the y-component of the solution on the slow invariant manifold (the *dashed line*), and its first-order approximation (the *dotted line*); (**a**) $\varepsilon = 0.1$, (**b**) $\varepsilon = 0.01$. This shows that the solution on the slow invariant manifold deviates significantly from the exact solution only in the initial layer

we have the exact solution in the form

$$y(t, \varepsilon) = \frac{y_0 e^{-t/\varepsilon}}{1 + y_0 - y_0 e^{-t/\varepsilon}},$$

$$x(t, \varepsilon) = x_0 + \varepsilon \ln (1 + y_0) + \varepsilon \ln \left(1 - \frac{y_0}{1 + y_0} e^{-t/\varepsilon} \right)$$

$$= x_0 + \varepsilon \ln \left(1 + y_0 - y_0 e^{-t/\varepsilon} \right), \tag{2.25}$$

when $y_0 > -1$ and, therefore, the argument of ln is nonnegative. It is clear that the y-component of the solution tends to the trajectory $y \equiv 0$ as $t \to \infty$. This means that the trajectory $y \equiv 0$ can be considered as an attractive slow invariant manifold, see Fig. 2.5. In this case the flow on the attractive slow invariant manifold is then described by the equation

$$\dot{x} = 0,$$

i.e. $x(t, \varepsilon) = const$, and this constant is the limit of the x-component of the exact solution $x(t, \varepsilon) \to x_0 + \varepsilon \ln (1 + y_0)$, which is the corresponding component of the solution on the attractive slow invariant manifold, as $t \to \infty$, see (2.25) and Fig. 2.6.

If $y_0 = -1$, the exact solution becomes

$$x = x_0 - t, \quad y \equiv -1,$$

and this means that $y = -1$ is also a slow invariant manifold.

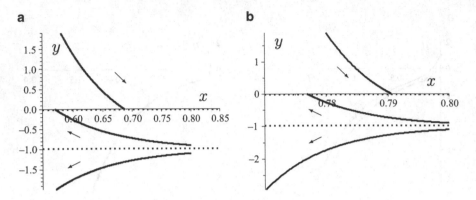

Fig. 2.5 Example 3: the slow invariant manifolds (the *dotted lines*) and the trajectories (the *solid lines*); (**a**) $\varepsilon = 0.1$, the initial points: $x(0) = 0.58$ and $y(0) = 1.9$, $x(0) = 0.8$ and $y(0) = -0.9$, $x(0) = 0.8$ and $y(0) = -1.1$; (**b**) $\varepsilon = 0.01$, the initial points: $x(0) = 0.78$ and $y(0) = 1.9$, $x(0) = 0.8$ and $y(0) = -0.9$, $x(0) = 0.8$ and $y(0) = -1.1$. The *arrows* indicate increasing time. So $y = 0$ is attractive and $y = -1$ is repulsive

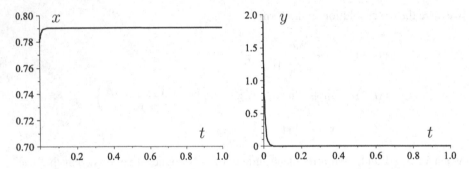

Fig. 2.6 Example 3: the x- and y-components of the exact solution as functions of t with $x_0 = 0.78$, $y_0 = 1.9$ and $\varepsilon = 0.01$

Consider the case $y_0 < 0$. Note that values $-1 < y_0 < 0$ fit in both cases ($y_0 > -1$ and $y_0 < 0$) with equal facility. On writing the y equation as

$$\left(-\frac{1}{y} + \frac{1}{1+y}\right) dy = \frac{dt}{\varepsilon},$$

the exact solution can be represented as

$$x(t, \varepsilon) = x_0 - t + \varepsilon \ln (-y_0) + \varepsilon \ln \left(1 - \frac{1 + y_0}{y_0} e^{t/\varepsilon}\right)$$

$$= x_0 - t + \varepsilon \ln \left((1 + y_0)e^{t/\varepsilon} - y_0\right),$$

$$y(t, \varepsilon) = -1 + \frac{(1 + y_0)e^{t/\varepsilon}}{(1 + y_0)e^{t/\varepsilon} - y_0}.$$

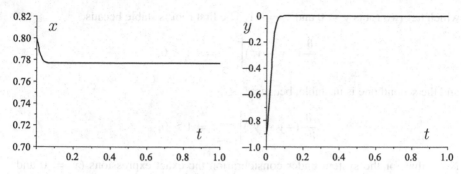

Fig. 2.7 Example 3: the x- and y-components of the exact solution as functions of t with $x_0 = 0.8$, $y_0 = -0.9$ and $\varepsilon = 0.01$. For both sets of initial conditions the exact solution, with $\varepsilon = 0.01$, tends rapidly to the solution $y(t) \equiv 0$

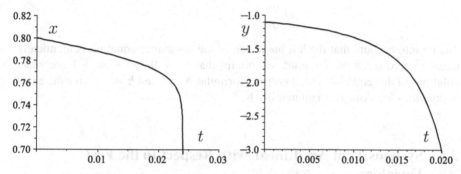

Fig. 2.8 Example 3: the x- and y-components of the solution as functions of t with $x_0 = 0.8$, $y_0 = -1.1$ and $\varepsilon = 0.01$

If $-1 < y_0 < 0$, than the y-component of the solution tends to $y \equiv 0$ as $t \rightarrow \infty$, see Fig. 2.7. If $y_0 < -1$, than this component tends to $-\infty$ as t changes from 0 to $t = \varepsilon \ln \frac{y_0}{1+y_0}$, where $(1 + y_0)e^{t/\varepsilon} - y_0 = 0$, see Fig. 2.8. Hence the trajectory $y \equiv -1$ is a repulsive slow invariant manifold. The flow on this repulsive slow invariant manifold is described by the equation

$$\dot{x} = -1.$$

Thus, the system under consideration possesses the attractive slow invariant manifold $y = 0$ and repulsive slow invariant manifold $y = -1$, see Fig. 2.5. This example illustrates the possibility of the coexistence of several slow integral manifolds.

If we use the formal procedure described, we begin from the equation for the slow curve

$$0 = -y - y^2,$$

which has two roots $y = 0$ and $y = -1$. The first root is stable because

$$\frac{\partial}{\partial y}(-y - y^2)\bigg|_{y=0} = -1 < 0,$$

and the second one is unstable, because

$$\frac{\partial}{\partial y}(-y - y^2)\bigg|_{y=-1} = 1 > 0.$$

Note, that for the system under consideration the exact expressions ($y = 0$ and $y = -1$) for the slow invariant manifolds coincide with their zero approximations, since here $f = y$, $g = -y - y^2$ and the invariance equation (2.4) for h is

$$\varepsilon h \frac{dh}{dx} = -h(1 + h).$$

Taking into account that the left hand side of the invariance equation is identically equal to zero for $h = $ constant, we obtain that $h = 0$ and $h = -1$ are exact solutions of this equation and, therefore, formulae $h = 0$ and $h = -1$ give the exact expressions for slow invariant manifolds.

2.5 Systems That Are Linear with Respect to the Fast Variables

Now we turn to systems that are linear with respect to the fast variable y, and consider the following equations

$$\dot{x} = \zeta(x, t, \varepsilon) + F(x, t, \varepsilon)y,$$
$$\varepsilon \dot{y} = \xi(x, t, \varepsilon) + G(x, t, \varepsilon)y, \tag{2.26}$$

where the righthand sides are linear with respect to the fast variable y. Such systems are typical of enzyme kinetics [44].

We assume that the eigenvalues $\lambda_i(x, t)$ of the matrix $G(x, t, 0)$ have the property $Re\lambda_i(x, t) \le -2\gamma < 0$, in $t \in \mathbb{R}, x \in \mathbb{R}^m$, and that the matrix- and vector-functions ζ, ξ, F and G are continuous and bounded as well as their partial derivatives with respect to the arguments $t \in \mathbb{R}, x \in \mathbb{R}^m, \varepsilon \in [0, \varepsilon_0]$. When these assumptions hold, the system (2.26) has a slow integral manifold

$$y = h(x, t, \varepsilon) = \phi(x, t) + \varepsilon h_1(x, t) + \dots.$$

On noting that

$$\frac{dy}{dt} = \frac{\partial h}{\partial t} + \frac{\partial h}{\partial x}(\zeta + Fh),$$

or using the first of (2.26), the functions h_i can be derived from the second of (2.26)

$$\varepsilon\frac{\partial h}{\partial t} + \varepsilon\frac{\partial h}{\partial x}(\zeta + Fh) = \xi + Gh.$$

Assume $\zeta = \sum\limits_{i=0}^{k} \varepsilon^i \zeta_i(x,t) + O(\varepsilon^{k+1})$ and a similar representation holds for ξ, F and G. Then the following recurrent relations hold:

$$\phi = -G_0^{-1}\xi_0, \quad h_1 = G_0^{-1}[\frac{\partial\phi}{\partial t} + \frac{\partial\phi}{\partial x}(\zeta_0 + F_0\phi) - \xi_1 - G_1\phi],$$

$$h_2 = G_0^{-1}[\frac{\partial h_1}{\partial t} + \frac{\partial\phi}{\partial x}(\zeta_1 + F_0h_1 + F_1\phi) + \frac{\partial h_1}{\partial x}(\zeta_0 + F_0\phi) - \xi_2 - G_2\phi - G_1h_1],$$

$$h_i = G_0^{-1}\{\frac{\partial h_{i-1}}{\partial t} + \sum_{j=0}^{i-1}\frac{\partial h_j}{\partial x}[\zeta_{i-j-1} + \sum_{s=0}^{i-j-1}F_s h_{i-j-s-1}]$$

$$- \xi_i - \sum_{j=1}^{i}G_j h_{i-j}\}, \quad i = 3, \ldots, k. \tag{2.27}$$

In many applications the $o(\varepsilon)$ order terms may be neglected, and we may then restrict ourselves to the first order approximation of the function $h = \phi(x,t) + \varepsilon h_1(x,t)$.

Example 4. As an example we consider the motion of a pendulum in a viscous medium. The motion is described by Newton's second law with the following autonomous system of equations

$$\dot{x} = y,$$
$$\varepsilon\dot{y} = -y - \sin x. \tag{2.28}$$

Clearly the system is linear in the fast variable y but nonlinear in x. We construct the slow invariant manifold of the pendulum equation (2.28).

The invariant manifold of slow motions $y = h(x, \varepsilon)$ for this system is given in the following form:

$$y = h(x, \varepsilon) = \phi(x) + \varepsilon h_1(x) + \varepsilon^2 h_2(x) + o(\varepsilon^2),$$

where $h(x, \varepsilon)$ satisfies

$$\varepsilon h\frac{dh}{dx} = -h - \sin x.$$

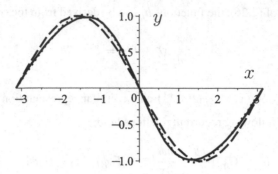

Fig. 2.9 Example 4: the slow curve $y = \phi(x) = -\sin x$ (the *dashed line*), the first-order $y = \phi(x) + \varepsilon h_1(x)$ (the *dotted line*) and the second-order $y = \phi(x) + \varepsilon h_1(x) + \varepsilon^2 h_2(x)$ (the *solid line*) approximations of the slow invariant manifold with $\varepsilon = 0.2$

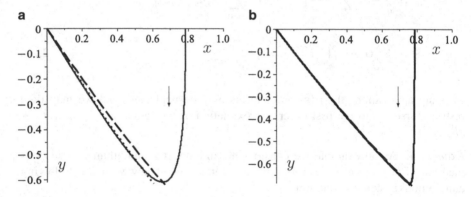

Fig. 2.10 Example 4: the trajectory (the *solid line*) of the solution with $x(0) = \pi/4$, $y(0) = 0$ (there is a boundary layer at the initial point $(\pi/4, 0)$), the zero-order $y = \phi(x)$ (the *dashed line*) and the first-order $y = \phi(x) + \varepsilon h_1(x)$ (the *dotted line*) approximations of the slow invariant manifold; (**a**) $\varepsilon = 0.1$, (**b**) $\varepsilon = 0.01$

Then

$$\phi = -\sin x, \ h_1 = -\frac{1}{2} \sin 2x, \ h_2 = -\sin x (\cos^2 x + \cos 2x).$$

The motions on this manifolds are described by the scalar equation

$$\dot{x} = h(x, \varepsilon) = -\sin x - \varepsilon \frac{1}{2} \sin 2x - \varepsilon^2 \sin x (\cos^2 x + \cos 2x) + O(\varepsilon^3).$$

Figures 2.9 and 2.10 demonstrate the results of the calculations.

Chapter 3
The Book of Numbers

Abstract In this chapter the first number in the title of a section denotes the dimension of the slow variable, the second one denotes the dimension of the fast variable. A series of examples, of increasing complexity, are given to illustrate the theoretical concepts. The main examples come from applications in enzyme kinetics. These examples illustrate the effectiveness of the order reduction method.

3.1 0+1

A number of examples involving scalar equations are given in this section to illustrate the concepts introduced in Chap. 2.

Consider the non-autonomous differential equation

$$\varepsilon \frac{dy}{dt} = g(y, t) \tag{3.1}$$

with scalar variable y, sufficiently smooth function g, positive small parameter ε and an initial condition

$$y = y_0 \quad \text{at} \quad t = t_0.$$

Equation (3.1) is of (0+1)-type, since the slow variable x is absent, so its dimension is equal to zero and the fast variable y is scalar, i.e. its dimension is equal to unity. Let $y = y(t, \varepsilon)$ be the solution of this initial value problem. Also let $y = \phi(t)$ be the solution (for simplicity we suppose that this solution is unique) of the corresponding degenerate equation $0 = g(y, t)$, obtained by setting $\varepsilon = 0$ in (3.1).

The question now arises of whether there is a relationship between $y(t, \varepsilon)$ and $\phi(t)$. If $y(t, \varepsilon) \to \phi(t)$ as $\varepsilon \to 0$, $\phi(t)$ is *stable* (or *attractive*); if $y(t, \varepsilon)$ moves away rapidly from $\phi(t)$ as $\varepsilon \to 0$, $\phi(t)$ is *unstable* (or *repulsive*). In order to compare the functions $y(t, \varepsilon)$ and $\phi(t)$ for small values of ε, consider the function

$$B(t) = \frac{\partial g(y, t)}{\partial y} \quad \text{on} \quad y = \phi(t), \quad \text{i.e.} \quad B(t) = \frac{\partial g(y, t)}{\partial y}\Big|_{y=\phi(t)}.$$

© Springer International Publishing Switzerland 2014   43
E. Shchepakina et al., *Singular Perturbations*, Lecture Notes in Mathematics 2114,
DOI 10.1007/978-3-319-09570-7_3

Sufficient conditions for the stability (instability) of $\phi(t)$ are [111]

◇ If $B(t) < 0$ then the solution of the degenerate equation, $y = \phi(t)$, is stable;
◇ If $B(t) > 0$ then the solution of the degenerate equation, $y = \phi(t)$, is unstable.

The proof of this fact for a special case can be found in Chap. 9 (see Sect. 9.1).

To find an approximate solution, playing the role of the slow integral manifold of (3.1), we use the form

$$y = \varphi(t, \varepsilon) = \phi(t) + \varepsilon\varphi_1 + \varepsilon^2\varphi_2 + \ldots, \tag{3.2}$$

which indicates that we are still ignoring terms of order $O(\varepsilon^3)$; the cutoff could be taken at any power of ε, but the more terms that are retained the longer the calculations become. The first step is to substitute (3.2) into (3.1), obtaining

$$\varepsilon(\dot{\phi}(t) + \varepsilon\dot{\varphi}_1 + \varepsilon^2\dot{\varphi}_2 + \ldots) = g(\phi(t) + \varepsilon\varphi_1 + \varepsilon^2\varphi_2 + \ldots, t). \tag{3.3}$$

The next step is to expand the r.h.s. of this equality in powers of ε by a Taylor series. If we are still reasoning formally, there is no need to pause over the justification of this step. It is convenient to have a name for the r.h.s. of (3.3) regarded as a function of ε, so we set $p(\varepsilon) = g(\phi(t) + \varepsilon\varphi_1 + \varepsilon^2\varphi_2 + \ldots, t)$. Expanding in powers of ε, ignoring terms at $O(\varepsilon^3)$, and noting $p(0) = g(\phi(t)) = 0$, leads to

$$p(0) + \varepsilon p'(0) + \frac{1}{2}\varepsilon^2 p''(0)$$

$$= \varepsilon g_y(\phi(t), t)\varphi_1 + \frac{1}{2}\varepsilon^2\left(g_{yy}(\phi(t), t)\varphi_1^2 + 2g_y(\phi(t), t)\varphi_2\right)$$

$$= \varepsilon B(t)(\varphi_1 + \varepsilon\varphi_2) + \frac{1}{2}\varepsilon^2 g_{yy}(\phi(t), t)\varphi_1^2.$$

Thus, we obtain from (3.3)

$$\varepsilon(\dot{\phi}(t) + \varepsilon\dot{\varphi}_1 + \varepsilon^2\dot{\varphi}_2 + \ldots) = \varepsilon B(t)(\varphi_1 + \varepsilon\varphi_2) + \frac{1}{2}\varepsilon^2 g_{yy}(\phi(t), t)\varphi_1^2,$$

and φ_1, φ_2 must satisfy

$$\dot{\phi}(t) = B(t)\varphi_1,$$

$$\dot{\varphi}_1 = B(t)\varphi_2 + \frac{1}{2}g_{yy}(\phi(t), t)\varphi_1^2.$$

As a result, the functions φ_1, φ_2 are given by

$$\varphi_1 = \dot{\phi}(t)/B(t),$$

$$\varphi_2 = [\dot{\varphi}_1 - \frac{1}{2}g_{yy}(\phi(t), t)\varphi_1^2]/B(t),$$

and $\phi(t)$ is defined by $g(\phi(t), t) = 0$.

Thus, the slow motion of (3.1) is described by the formula (3.2) for $y = \varphi(t, \varepsilon)$, where $\phi, \varphi_1, \varphi_2$ are given above.

If the degenerate equation has several solutions $y = \phi_i(t)$, $i = 1, 2 \ldots, k$ it is necessary to verify the stability of each solution. Then the behaviour of the solution $y = y(t, \varepsilon)$, as $\varepsilon \to 0$, depends on initial point (t_0, y_0).

Example 5. Consider a scalar equation

$$\varepsilon \dot{y} = -(3t^2 + 1)(y + \sin(t^3 + t)) \tag{3.4}$$

with the initial condition $y(t_0) = y_0$.

The degenerate equation is

$$g(y, t) = -(3t^2 + 1)(y + \sin(t^3 + t)) = 0$$

and has a stable solution $\phi(t) = -\sin(t^3 + t)$, since $B(t) = \frac{\partial g}{\partial y}\big|_{y=\phi(t)} = -(3t^2 + 1) < 0$.

The approximate stable slow integral manifold of the above equation is given by the perturbation expansion

$$y(t, \varepsilon) = \phi(t) + \varepsilon\varphi_1 + \varepsilon^2\varphi_2 + \ldots,$$

where

$$\phi(t) = -\sin(t^3 + t),$$
$$\varphi_1 = \dot{\phi}(t)/B(t) = \cos(t^3 + t),$$
$$\varphi_2 = (\dot{\varphi}_1 - \frac{1}{2}g_{yy}(\phi(t), t)\varphi_1^2)/B(t) = \sin(t^3 + t).$$

The approximate slow integral manifold is therefore

$$\phi(t) + \varepsilon\varphi_1 + \varepsilon^2\varphi_2 + o(\varepsilon^2) = -\sin(t^3 + t) + \varepsilon\cos(t^3 + t) + \varepsilon^2\sin(t^3 + t) + o(\varepsilon^2).$$

The exact solution of the original equation (3.4) is easily found using the change of variable $\eta = t^3 + t$, and the result is

$$y(t, \varepsilon) = \left[y_0 + \frac{1}{1 + \varepsilon^2}\left(\sin(t_0^3 + t_0) - \varepsilon\cos(t_0^3 + t_0)\right) \right] e^{(t_0^3 + t_0 - t^3 - t)/\varepsilon}$$
$$- \frac{1}{1 + \varepsilon^2}\left[\sin(t^3 + t) - \varepsilon\cos(t^3 + t) \right].$$

Fig. 3.1 Example 5: the slow integral manifold (the *dotted line*), the slow curve $y = \phi(t)$ (the *dashed line*) and two trajectories (the *solid lines*) corresponding to different initial conditions and approaching the slow integral manifold; $\varepsilon = 0.1$. The *arrows* indicate increasing time

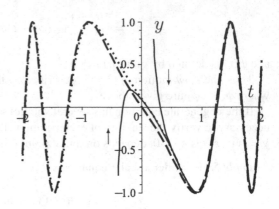

Equation (3.4) has the exact slow integral manifold

$$y(t,\varepsilon) = \frac{1}{1 + \varepsilon^2} \left(\varepsilon \cos(t^3 + t) - \sin(t^3 + t) \right)$$

which is attractive as $t \to \infty$, and this agrees with the results of the perturbation scheme with an error $o(\varepsilon^2)$.

The exact solution contains the effect of the initial condition at $t = t_0$, and this decays rapidly for $t > t_0$ to leave only the attractive slow integral manifold. The latter does not contain the initial condition: in some sense the "initial layer" is missing and doesn't affect the steady state solution. This is the essence of Tikhonov's theorem, see (1.34), (1.35).

In Fig. 3.1 we show the slow integral manifold, the slow curve $y = \phi(t) = -\sin(t^3 + t)$ and some trajectories with $\varepsilon = 0.1$.

Example 6. Consider a non-autonomous scalar equation

$$\varepsilon \dot{y} = (3t^2 + 1)(y + \sin(t^3 + t))$$

with the initial condition $y(t_0) = y_0$.

The degenerate equation

$$(3t^2 + 1)(y + \sin(t^3 + t)) = 0$$

has an unstable solution $\phi(t) = -\sin(t^3 + t)$ due to $B(t) = 3t^2 + 1 > 0$.

The approximate unstable slow integral manifold of this equation is given by the perturbation expansion

$$y(t,\varepsilon) = \phi(t) + \varepsilon \varphi_1 + \varepsilon^2 \varphi_2 + \ldots,$$

where

$$\phi(t) = -\sin(t^3 + t),$$
$$\varphi_1 = \dot{\phi}(t)/B(t) = -\cos(t^3 + t),$$
$$\varphi_2 = (\dot{\varphi}_1 - \frac{1}{2}\varepsilon^2 g_{yy}(\phi(t), t)\varphi_1^2)/B(t) = \sin(t^3 + t).$$

The exact solution is

$$y(t, \varepsilon) = \left\{ y_0 + \frac{1}{1 + \varepsilon^2} \left[\varepsilon \cos(t_0^3 + t_0) + \sin(t_0^3 + t_0) \right] \right\} e^{(t^3 + t - t_0^3 - t_0)/\varepsilon}$$
$$- \frac{1}{1 + \varepsilon^2} \left(\varepsilon \cos(t^3 + t) + \sin(t^3 + t) \right).$$

The equation under consideration thus possesses the exact repulsive slow integral manifold

$$y(t, \varepsilon) = -\frac{1}{1 + \varepsilon^2} \left(\varepsilon \cos(t^3 + t) + \sin(t^3 + t) \right),$$

and again the perturbation procedure gives the approximate result with an error at $o(\varepsilon^2)$ for $t > t_0$, but does not contain the effect of the initial condition at $t = t_0$.

Figure 3.2 shows the slow integral manifold, the slow curve and two trajectories corresponding to different initial conditions, with $\varepsilon = 0.1$.

Example 7. Consider the nonlinear autonomous initial value problem

$$\varepsilon \dot{y} = y(y^2 - 1), \quad y(0) = y_0. \tag{3.5}$$

The degenerate equation $g(y) = y(y^2 - 1) = 0$ has three solutions: $\phi_1(t) = -1$, $\phi_2(t) = 0$ and $\phi_3(t) = +1$. Then

$$B(t) = \frac{\partial g}{\partial y}\bigg|_{y = \phi(t)}$$

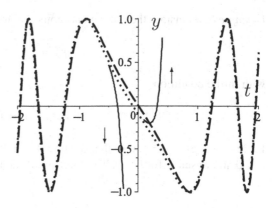

Fig. 3.2 Example 6: the slow integral manifold (the *dotted line*), the slow curve (the *dashed line*) and two trajectories leaving the repulsive slow integral manifold (the *solid lines*); $\varepsilon = 0.1$. The *arrows* indicate increasing time

yields

$$B(t) = (3y^2 - 1)\big|_{y=\phi_{1,3}(t)} = 3 - 1 = 2 > 0, \quad B(t) = (3y^2 - 1)\big|_{y=\phi_2(t)} = -1 < 0.$$

Therefore, $\phi_2(t)$ is stable, $\phi_1(t)$ and $\phi_3(t)$ are unstable for all t.

The exact solution of (3.5) can be found by separation of variables, and the solution is found by solving

$$\int \frac{dt}{\varepsilon} = \int \frac{dy}{y(y^2 - 1)} = \int \left(-\frac{1}{y} + \frac{1}{2(y-1)} + \frac{1}{2(y+1)} \right) dy$$

for y. This gives

$$\frac{t}{\varepsilon} = \ln \left(\frac{|y^2 - 1|}{y^2} \right)^{1/2} + C$$

with the integration constant C. Taking into account the initial conditions we obtain the solution in the form

$$y(t) = y_0 e^{-t/\varepsilon} \left(1 - y_0^2 + y_0^2 e^{-2t/\varepsilon} \right)^{-1/2}.$$

Then

$$y(t) \to 0 \text{ as } t \to +\infty, \quad \text{if } y_0^2 < 1,$$

$$y(t) \to +\infty \text{ as } t \to t^*, t^* = \frac{\varepsilon}{2} \ln \frac{y_0^2}{y_0^2 - 1}, \quad \text{if } y_0 > 1,$$

and

$$y(t) \to -\infty \text{ as } t \to t^*, \quad \text{if } y_0 < 1.$$

Thus the solution $y(t) \equiv 0$ is the stable slow integral manifold, $y(t) \equiv -1$ and $y(t) \equiv +1$ are unstable slow integral manifolds, see Fig. 3.3.

Example 8. Consider the non-autonomous scalar equation

$$\varepsilon \dot{y} = ty \tag{3.6}$$

with initial condition

$$y = y_0 \quad \text{at} \quad t = t_0.$$

In this case, $g(y, t) = ty$ so that $\phi(t) \equiv 0$ and $B(t) = t$. Hence, $\phi(t)$ is stable for $t < 0$ and unstable for $t > 0$. The solution of the initial value problem for (3.6) is

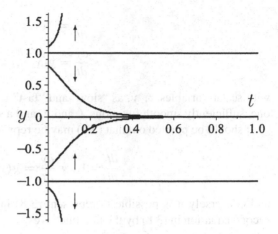

Fig. 3.3 Example 7: the solutions of (3.5) with different initial values and $\varepsilon = 0.1$. The *arrows* indicate increasing time

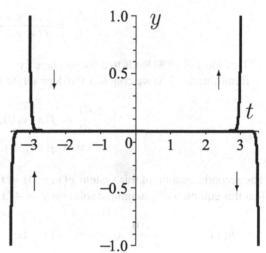

Fig. 3.4 Example 8: the solutions of (3.6) with different initial values and $\varepsilon = 0.1$. The *arrows* indicate increasing time

$$y(t) = y_0 e^{(t^2 - t_0^2)/2\varepsilon}. \tag{3.7}$$

Note that the solution $y(t) \equiv 0$, corresponding to $y_0 = 0$, plays the role of the slow integral manifold, which is attractive for $t < 0$ and repulsive for $t > 0$, see Fig. 3.4.

3.2 1+1

3.2.1 Theoretical Background

Moving to the next level of the complexity, we consider a system of two ordinary autonomous differential equations

$$\frac{dx}{dt} = f(x, y, \varepsilon),$$

$$\varepsilon \frac{dy}{dt} = g(x, y, \varepsilon),$$

(3.8)

with scalar variables x, y, as "slow" and "fast" respectively, both with dimension one, sufficiently smooth functions f and g, and a small positive parameter ε.

It should be pointed out that (3.1) may be represented in the form (3.8):

$$\frac{dt}{dt} = 1, \quad \varepsilon \frac{dy}{dt} = g(y, t),$$

and conversely it is possible to represent (3.8) in the form (3.1). Division of the second equation in (3.8) by the first one gives

$$\varepsilon \frac{dy}{dx} = \frac{g(x, y, \varepsilon)}{f(x, y, \varepsilon)}.$$

This form of (3.8) leads to a phase-plane $(y - x)$ analysis.

Returning to (3.8) we consider the degenerate system

$$\frac{dx}{dt} = f(x, y, 0),$$

$$0 = g(x, y, 0).$$

The second equation of this system $g(x, y, 0) = 0$ describes a *slow curve*. Suppose that this equation has an unique solution $y = \phi(x)$. Introduce the function

$$B(x) = \frac{\partial g(x, y, 0)}{\partial y} \quad \text{on} \ \ y = \phi(x), \quad \text{i.e.} \quad B(x) = \frac{\partial g(x, y, 0)}{\partial y}\Big|_{y=\phi(x)}.$$

Sufficient conditions for the stability (instability) of $\phi(x)$ are identical to those for (3.1)

◇ If $B(x) < 0$ then the solution of the degenerate equation, $y = \phi(x)$, is stable;
◇ If $B(x) > 0$ then the solution of the degenerate equation, $y = \phi(x)$, is unstable.

If the degenerate equation has several solutions $y = \phi_i(x)$, $i = 1, 2 \dots, k$ it is necessary to check each solution for stability. Then the behavior of the solution

$$x = x(t, \varepsilon), \quad x(t_0, \varepsilon) = x_0, \quad y = y(t, \varepsilon), \quad y(t_0, \varepsilon) = y_0$$

as $\varepsilon \to 0$ depends on initial point (x_0, y_0), i.e., does it or does it not lie within the domain of attraction of a stable slow curve $y = \phi(x)$.

3.2.1.1 Asymptotic Expansions

To obtain the asymptotic expansion for a one-dimensional slow invariant manifold

$$y = h(x, \varepsilon) = \phi(x) + \varepsilon h_1(x) + \cdots + \varepsilon^k h_k(x) + \dots \,,$$

we substitute this formal expansion into the equation

$$\varepsilon \frac{dy}{dx} = \frac{g(x, y, \varepsilon)}{f(x, y, \varepsilon)},$$

or, in more convenient form, into the invariance equation (2.4)

$$\varepsilon \frac{dh}{dx} f(x, h(x, \varepsilon), \varepsilon) = g(x, h, \varepsilon).$$

We could use the general formulas from Sect. 2.3, but instead we will calculate the asymptotic expansion in the form

$$h(x, \varepsilon) = \phi(x) + \varepsilon h_1(x) + O(\varepsilon^2).$$

Thus, we obtain the relationship

$$\varepsilon \frac{d\phi}{dx} f(x, \phi(x), 0) + O(\varepsilon^2) = g(x, \phi(x) + \varepsilon h_1(x) + O(\varepsilon^2), \varepsilon).$$

We use the formal asymptotic representations

$$g(x, \phi(x) + \varepsilon h_1(x) + O(\varepsilon^2), \varepsilon) = B(x)(\varepsilon h_1 + O(\varepsilon^2)) + \varepsilon g_1(x, \phi) + O(\varepsilon^2),$$

on taking into account

$$g(x, \phi(x), 0) = 0,$$

where the function $B(x) \equiv (\partial g / \partial y)(x, \phi(x), 0)$, and

$$g_1(x, \phi) \equiv (\partial g / \partial \varepsilon)(x, \phi(x), 0).$$

Substituting these formal representations into the invariance equation and equating powers of ε, we obtain

$$h_1 = \left[\frac{d\phi}{dx} f_0 - g_1 \right] / B,$$

where $f_0 = f(x, \phi(x), 0)$.

The slow motions of (3.8) are now described by the equations

$$\frac{dx}{dt} = f(x, \phi(x) + \varepsilon h_1(x) + \dots, \varepsilon),$$

$$y = h(x, \varepsilon) = \phi(x) + \varepsilon h_1(x) + \dots,$$

where $h_1(x)$ is given above.

3.2.2 Michaelis–Menten Kinetics

We use the dimensionless 2D model of Michaelis–Menten kinetics of enzyme action, (see [118]), to illustrate the $(1 + 1)$-case (i.e. $\dim x = 1$, $\dim y = 1$).

The Michaelis–Menten mechanism is written

$$E + S \underset{k_{-1}}{\overset{k_1}{\rightleftharpoons}} ES \overset{k_2}{\rightarrow} E + P. \tag{3.9}$$

In the scheme (3.9), E represents an enzyme, S a substrate, ES an enzyme–substrate complex, and P a product. We write $C = ES$ for the intermediate complex. Also, concentrations are indicated by lower case letters, i.e., x is a concentration of X, and the time derivative dx/dt by $\dot{x}$. In this notation the system of differential equations for the scheme (3.9) is

$$\dot{e} = -k_1 es + k_{-1} c + k_2 c, \tag{3.10}$$

$$\dot{s} = -k_1 es + k_{-1} c, \tag{3.11}$$

$$\dot{c} = k_1 es - k_{-1} c - k_2 c, \tag{3.12}$$

$$\dot{p} = k_2 c. \tag{3.13}$$

The relevant initial conditions for (3.10)–(3.13) are that the concentrations of the substrate s and the enzyme e are given and non-zero and those of the complex c and product p are zero, that is,

$$s(0) = s_0 \neq 0, \quad e(0) = e_0 \neq 0, \quad c(0) = p(0) = 0. \tag{3.14}$$

The ultimate purpose is to find the steady state form of the substrate S and the concentration of the intermediate complex ES.

This system immediately yields two constants of the motion: "total enzyme" and "total substrate." Adding Eqs. (3.10) and (3.12) gives

$$d(e + c)/dt = 0,$$

so that

$$e + c = e_0, \tag{3.15}$$

the total enzyme concentration, since $c(0) = 0$. Adding Eqs. (3.11), (3.12) and (3.13) gives

$$d(s + c + p)/dt = 0,$$

so integration gives

$$s + c + p = s_0 \tag{3.16}$$

the total substrate concentration, since $c(0) = p(0) = 0$. Using Eq. (3.15), e may be eliminated from Eqs. (3.11) and (3.12), to give the closed, coupled pair of equations

$$\dot{s} = -k_1(e_0 - c)s + k_{-1}c, \tag{3.17}$$

$$\dot{c} = k_1(e_0 - c)s - k_{-1}c - k_2c. \tag{3.18}$$

We introduce the nondimensional quantities

$$\tau = k_1 e_0 t, \quad \lambda = \frac{k_2}{k_1 s_0}, \quad \kappa = \frac{k_{-1} + k_2}{k_1 s_0},$$

$$x(\tau) = \frac{s(t)}{s_0}, \quad y(\tau) = \frac{c(t)}{e_0}, \quad \varepsilon = \frac{e_0}{s_0}, \tag{3.19}$$

where e_0 and s_0 are the initial enzyme and substrate concentrations in (3.14). All of x, y, τ, λ, κ and ε in (3.19) are dimensionless variables and parameters independent of the system of units used. Substituting (3.19) into the system (3.17), (3.18) with the initial conditions from (3.14) they become the following nondimensional system for $x(\tau)$ and $y(\tau)$:

$$\frac{dx}{d\tau} = -x + (x + \kappa - \lambda)y, \tag{3.20}$$

$$\varepsilon \frac{dy}{d\tau} = x - (x + \kappa)y, \tag{3.21}$$

with initial conditions

$$x(0) = 1, \quad y(0) = 0. \tag{3.22}$$

In most biological situations the ratio of the initial enzyme to the initial substrate is small, that is $\varepsilon = e_0/s_0 \ll 1$, and so (3.20)–(3.22) is a singular perturbation problem.

The degenerate system is

$$\frac{dx}{d\tau} = -x + (x + \kappa - \lambda)y,$$

$$0 = x - (x + \kappa)y.$$

The last equation has the solution

$$y = \phi(x) = \frac{x}{x + \kappa}.$$

To find the slow invariant manifold of (3.20), (3.21) in the form of an asymptotic expansion

$$y = h(x, \varepsilon) = \phi(x) + \varepsilon h_1(x) + \dots \tag{3.23}$$

we substitute (3.23) into (3.21) and use (3.20) to get the invariance equation

$$\varepsilon \frac{dh(x, \varepsilon)}{dx}[-x + (x + \kappa - \lambda)h(x, \varepsilon)] = x - (x + \kappa)h(x, \varepsilon).$$

Then

$$\varepsilon (\phi'(x) + \varepsilon h_1'(x) + \dots)[-x + (x + \kappa - \lambda)(\phi(x) + \varepsilon h_1(x) + \dots)]$$
$$= x - (x + \kappa)(\phi(x) + \varepsilon h_1(x) + \varepsilon^2 h_2(x) + \dots). \tag{3.24}$$

Equating the coefficients of the first power of ε in (3.24), and noting that

$$x - (x + \kappa)\phi(x) = 0 \text{ or } \phi(x) = \frac{x}{x + \kappa},$$

we obtain

$$\phi'(x)[-x + (x + \kappa - \lambda)\phi(x)] = -(x + \kappa)h_1(x).$$

From this and

$$\phi'(x) = \frac{\kappa}{(x + \kappa)^2}$$

we calculate

$$h_1(x) = \frac{\lambda \kappa x}{(x + \kappa)^4}.$$

Thus, the slow invariant manifold is

$$y = \frac{x}{x + \kappa} + \varepsilon \frac{\lambda \kappa x}{(x + \kappa)^4} + O(\varepsilon^2), \tag{3.25}$$

and the flow on it is described by

$$\frac{dx}{d\tau} = \frac{-\lambda x}{x + \kappa} + \varepsilon \frac{\lambda \kappa x (x + \kappa - \lambda)}{(x + \kappa)^4} + O(\varepsilon^2). \tag{3.26}$$

This slow invariant manifold is attractive since

$$\frac{\partial g}{\partial y} = \frac{\partial}{\partial y}(x - (x + \kappa)y) = -(x + \kappa) < 0,$$

and thus there is the unique stable equilibrium $x = 0$ to Eq. (3.26) on this invariant manifold.

In Fig. 3.5 we can see that the trajectory of (3.20)–(3.22) approaches the slow invariant manifold very rapidly from the initial point $x(0) = 1$, $y(0) = 0$, and then tends to the origin which is the equilibrium of (3.20)–(3.22) along the manifold as $t \to \infty$.

The zero approximation

$$\frac{dx}{d\tau} = -x + (x + \kappa - \lambda)y = -\frac{\lambda x}{x + \kappa}, \tag{3.27}$$

$$y = \phi(x) = \frac{x}{x + \kappa},$$

when put in dimensional variables, gives the well-known *Michaelis–Menten kinetic law*:

$$\frac{ds}{dt} = e_0 \frac{-k_2 s}{K + s}, \quad c = e_0 \frac{s}{K + s}, \quad K = \frac{k_{-1} + k_2}{k_1}. \tag{3.28}$$

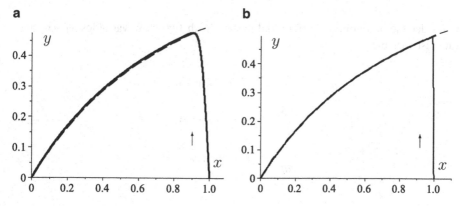

Fig. 3.5 The trajectory (the *solid line*) of (3.20)–(3.22) and the slow curve (the *dashed line*) for $\kappa = 1$, $\lambda = 0.5$, and (**a**) $\varepsilon = 0.1$, (**b**) $\varepsilon = 0.01$. The *arrows* indicate increasing time

The first order approximation (3.25), (3.26) is as follows in dimensional terms

$$\frac{ds}{dt} = e_0 \frac{-k_2 s}{K + s} + e_0^2 \frac{(k_1 s + k_{-1}) K k_2 s}{k_1 (K + s)^4},$$

$$c = e_0 \frac{s}{K + s} + e_0^2 \frac{K k_2 s}{k_1 (K + s)^4}.$$

These relationships may be called *the generalized Michaelis–Menten kinetic law.*

Since the slow invariant manifold is attractive and $0 < \varepsilon \ll 1$, the trajectory lands on the manifold very quickly after the initial instant and then flows to the origin. Thus an "initial layer" perturbation scheme is of little interest to the long-term state. However, the reader may wish to do a matched asymptotic expansion to find a uniformly valid solution for $t \geq 0$, see [119] for a detailed discussion of this problem. R.S. Johnson [80, p. 263], uses a multiple-scale expansion to solve the problem.

Note that Eq. (3.27) may be solved exactly:

$$x(t) + \kappa \ln x(t) = 1 - \lambda t$$

on using $x(0) = 1$.

3.3   2+1

We now consider the case where the slow variable has dimension 2, while the fast variable has dimension 1.

3.3.1   *Theoretical Background*

Consider the autonomous differential system with two slow variables x_1, x_2 and one fast variable y

$$\frac{dx_1}{dt} = f_1(x_1, x_2, y, \varepsilon),$$

$$\frac{dx_2}{dt} = f_2(x_1, x_2, y, \varepsilon), \qquad (3.29)$$

$$\varepsilon \frac{dy}{dt} = g(x_1, x_2, y, \varepsilon),$$

where ε is a small positive parameter. The corresponding degenerate system is

$$\frac{dx_1}{dt} = f_1(x_1, x_2, y, 0),$$

$$\frac{dx_2}{dt} = f_2(x_1, x_2, y, 0),$$

$$0 = g(x_1, x_2, y, 0).$$

The last equation describes the two-dimensional slow surface in implicit form.
Let this equation have the solution (i.e., the slow surface has the form)

$$y = \phi(x_1, x_2).$$

The slow surface $y = \phi(x_1, x_2)$ is stable, if

$$B(x_1, x_2) = \frac{\partial g(x_1, x_2, y, 0)}{\partial y}\Big|_{y=\phi(x_1,x_2)}$$

is negative, and is unstable if $B(x_1, x_2) > 0$.

To obtain the asymptotic expansion for the two-dimensional slow invariant manifold

$$y = h(x_1, x_2, \varepsilon) = \phi(x_1, x_2) + \varepsilon h_1(x_1, x_2) + \dots ,$$

we substitute this formal expansion into the invariance equation for (3.29):

$$\varepsilon \frac{\partial h(x_1, x_2, \varepsilon)}{\partial x_1} f_1(x_1, x_2, h(x_1, x_2, \varepsilon), \varepsilon) + \varepsilon \frac{\partial h(x_1, x_2, \varepsilon)}{\partial x_2} f_2(x_1, x_2, h(x_1, x_2, \varepsilon), \varepsilon)$$
$$= g(x_1, x_2, h(x_1, x_2, \varepsilon), \varepsilon),$$

and obtain the relationship, on noting that $g(x_1, x_2, \phi(x_1, x_2), 0) = 0$,

$$\varepsilon \frac{\partial \phi(x_1, x_2)}{\partial x_1} f_1(x_1, x_2, \phi(x_1, x_2), 0) + \varepsilon \frac{\partial \phi(x_1, x_2)}{\partial x_2} f_2(x_1, x_2, \phi(x_1, x_2), 0) + O(\varepsilon^2)$$

$$= B(x_1, x_2)(\varepsilon h_1(x_1, x_2) + \dots) + \varepsilon \frac{\partial g}{\partial \varepsilon}(x_1, x_2, \phi(x_1, x_2), 0) + O(\varepsilon^2).$$

This implies, in particular, that

$$h_1(x_1, x_2) = \left[\frac{\partial \phi(x_1, x_2)}{\partial x_1} f_1(x_1, x_2, \phi(x_1, x_2), 0) + \frac{\partial \phi(x_1, x_2)}{\partial x_2} f_2(x_1, x_2, \phi(x_1, x_2), 0) \right.$$

$$\left. - \frac{\partial g}{\partial \varepsilon}(x_1, x_2, \phi(x_1, x_2), 0) \right] B^{-1}(x_1, x_2).$$

Thus, we can construct the slow invariant manifold

$$y = h(x_1, x_2, \varepsilon) = \phi(x_1, x_2) + \varepsilon h_1(x_1, x_2) + \dots ,$$

where $h_1(x_1, x_2)$ is given above. The motion of the system (3.29) on the slow invariant manifold is described by

$$\frac{dx_1}{dt} = f_1(x_1, x_2, h(x_1, x_2, \varepsilon), \varepsilon),$$

$$\frac{dx_2}{dt} = f_2(x_1, x_2, h(x_1, x_2, \varepsilon), \varepsilon),$$

and may be approximated by using the asymptotic expansion for $h(x_1, x_2, \varepsilon)$.

3.3.2   Bimolecular Reaction System

As an example of the above we consider the bimolecular reaction system [148]

$$S \to X,$$

$$2Y \underset{k_{-1}}{\overset{k_1}{\rightleftharpoons}} Z,$$

$$X + Z \overset{k_2}{\to} Y + Z,$$

$$Y \to P.$$

(3.30)

Here S and P denote substances with constant concentrations; k_1, k_{-1}, k_2 are positive parameters, k_{-1} is assumed to be "large". Under the assumptions of spatial homogeneity and mass–action kinetics and introducing the small parameter $\varepsilon = 1/k_{-1}$, the dynamic behavior is described by the differential equations in dimensionless form

$$\frac{dx_1}{dt} = 1 - k_2 x_1 y = f(x_1, z, y),$$

(3.31)

$$\varepsilon \frac{dz}{dt} = -2\varepsilon k_1 z^2 + 2y + \varepsilon(k_2 x_1 y - z) = g_1(x_1, z, y, \varepsilon),$$

(3.32)

$$\varepsilon \frac{dy}{dt} = \varepsilon k_1 z^2 - y = g_2(x_1, z, y, \varepsilon),$$

(3.33)

Setting $\varepsilon = 0$ we obtain: $2y = 0$, $y = 0$ from (3.32) and (3.33) and the corresponding Jacobian matrix

$$B = \left(\begin{array}{cc} \dfrac{\partial g_1}{\partial z} & \dfrac{\partial g_1}{\partial y} \\ \dfrac{\partial g_2}{\partial z} & \dfrac{\partial g_2}{\partial y} \end{array} \right) \Bigg|_{\varepsilon=0, y=0} = \left(\begin{array}{cc} 0 & 2 \\ 0 & -1 \end{array} \right)$$

is degenerate; i.e., $\det B = 0$. This means that the method of invariant manifolds cannot be applied directly to the system (3.31)–(3.33), see (2.6) in the assumption (II). However, introducing the new variable x_2 by

$$x_2 = z + 2y$$

leads to the system with nondegenerate matrix B

$$\frac{dx_1}{dt} = 1 - k_2 x_1 y = f_1(x_1, x_2, y, \varepsilon), \tag{3.34}$$

$$\frac{dx_2}{dt} = k_2 x_1 y - x_2 + 2y = f_2(x_1, x_2, y, \varepsilon), \tag{3.35}$$

$$\varepsilon \frac{dy}{dt} = -y + \varepsilon k_1 (x_2 - 2y)^2 = g(x_1, x_2, y, \varepsilon), \tag{3.36}$$

which is appropriate to the approach under consideration.

Now setting $\varepsilon = 0$ we obtain: $y = 0$, and the corresponding Jacobian matrix is $\dfrac{\partial g_2}{\partial y}(x_1, x_2, 0, 0) = B(x_1, x_2) = (-1)$ since $g(x_1, x_2, y, 0) = -y$. Thus, the system (3.34)–(3.36) has an attractive slow invariant manifold $y = h(x_1, x_2, \varepsilon)$, where $h(x_1, x_2, 0) = \phi(x_1, x_2) = 0$. Therefore, the asymptotic expansion is

$$y = h(x_1, x_2, \varepsilon) = \varepsilon h_1(x_1, x_2) + \varepsilon^2 h_2(x_1, x_2) + \dots . \tag{3.37}$$

The flow on this manifold is described by the following differential system

$$\frac{dx_1}{dt} = 1 - \varepsilon k_2 x_1 (h_1(x_1, x_2) + \varepsilon h_2(x_1, x_2) + \dots), \tag{3.38}$$

$$\frac{dx_2}{dt} = -x_2 + \varepsilon(2 + k_2 x_1)(h_1(x_1, x_2) + \varepsilon h_2(x_1, x_2) + \dots), \tag{3.39}$$

and the invariance equation for $h(x_1, x_2, \varepsilon)$ is

$$\varepsilon^2 \left(\frac{\partial h_1}{\partial x_1} + \dots \right) [1 - \varepsilon k_2 x_1 (h_1 + \dots)]$$

$$+ \varepsilon^2 \left(\frac{\partial h_1}{\partial x_2} + \dots \right) [-x_2 + \varepsilon(2 + k_2 x_1)(h_1 + \dots)] \tag{3.40}$$

$$= -\varepsilon h_1 - \varepsilon^2 h_2 - \dots + \varepsilon k_1 (x_2 - \varepsilon 2 h_1 - \dots)^2.$$

Equating powers of ε we obtain:

ε^1:

$$h_1 = k_1 x_2^2;$$

ε^2:

$$\frac{\partial h_1}{\partial x_1} \cdot 1 + \frac{\partial h_1}{\partial x_2}(-x_2) = -h_2 - 4k_1 x_2 h_1,$$

i.e.,

$$h_2 = 2k_1 x_2^2(1 - 2k_1 x_2).$$

Now we can write the second order approximation to the slow motion of (3.31)–(3.33)

$$\frac{dx_1}{dt} = 1 - \varepsilon k_1 k_2 x_1 x_2^2 + \varepsilon^2 2k_1 k_2 x_1 x_2^2(1 - 2k_1 x_2) + O(\varepsilon^3), \tag{3.41}$$

$$\frac{dx_2}{dt} = -x_2 + \varepsilon(2 + k_2 x_1)k_1 x_2^2 + \varepsilon^2(2 + k_2 x_1)2k_1 x_2^2(1 - 2k_1 x_2)$$

$$+ O(\varepsilon^3), \tag{3.42}$$

$$y = \varepsilon k_1 x_2^2 + \varepsilon^2 2k_1 x_2^2(1 - 2k_1 x_2) + O(\varepsilon^3), \tag{3.43}$$

$$z = x_2 - 2y. \tag{3.44}$$

The slow invariant manifold is given by Eqs. (3.43), (3.44), and the flow on this manifold is described by Eqs. (3.41), (3.42). The trajectory of the system (3.34)–(3.36) approaches the corresponding trajectory on the slow invariant manifold as $t \to \infty$. We will return to this model at the end of Sect. 5.2.

3.4   1+2

3.4.1   Theoretical Background

Consider the autonomous differential system with one slow variable x and two fast variables y_1, y_2

$$\frac{dx}{dt} = f(x, y_1, y_2, \varepsilon),$$

$$\varepsilon \frac{dy_1}{dt} = g_1(x, y_1, y_2, \varepsilon), \tag{3.45}$$

$$\varepsilon \frac{dy_2}{dt} = g_2(x, y_1, y_2, \varepsilon),$$

with a small positive parameter ε. The corresponding degenerate system is

$$\frac{dx}{dt} = f(x, y_1, y_2, 0),$$

$$0 = g_1(x, y_1, y_2, 0),$$

$$0 = g_2(x, y_1, y_2, 0).$$

The last two equations give a description of a one-dimensional slow manifold (slow curve). Suppose that these equations can be solved for y_1 and y_2, i.e.,

$$y_1 = \bar{\phi}(x), \quad y_2 = \bar{\bar{\phi}}(x).$$

Consider the Jacobian matrix

$$B = \begin{pmatrix} \dfrac{\partial g_1}{\partial y_1} & \dfrac{\partial g_1}{\partial y_2} \\ \dfrac{\partial g_2}{\partial y_1} & \dfrac{\partial g_2}{\partial y_2} \end{pmatrix}$$

along the slow curve i.e.,

$$B = B(x) = \begin{pmatrix} \dfrac{\partial g_1}{\partial y_1} & \dfrac{\partial g_1}{\partial y_2} \\ \dfrac{\partial g_2}{\partial y_1} & \dfrac{\partial g_2}{\partial y_2} \end{pmatrix}\Bigg|_{y_1=\bar{\phi}(x),\ y_2=\bar{\bar{\phi}}(x),\ \varepsilon=0}.$$

If both two roots of the characteristic polynomial

$$\det(B(x) - \lambda \mathbf{I}) = 0,$$

where $\mathbf{I}$ is the identity matrix, have negative real parts, then the slow curve is stable. Let

$$b_{ij}(x) = \frac{\partial g_i}{\partial y_j}\Bigg|_{y_1=\bar{\phi}(x),\ y_2=\bar{\bar{\phi}}(x),\ \varepsilon=0}, \quad i, j = 1, 2,$$

then

$$\text{tr}\, B(x) = b_{11}(x) + b_{22}(x),$$

and

$$\det B(x) = b_{11}(x)b_{22}(x) - b_{12}(x)b_{21}(x),$$

where tr $B(x)$ is the trace and det $B(x)$ is the determinant of $B(x)$. The condition for the stability of the slow curve is the positivity of the coefficients of the quadratic characteristic polynomial, which are $-\operatorname{tr} B(x)$ and det $B(x)$. This fact can be checked directly as follows.

Using the representation of the roots of the quadratic polynomial

$$\lambda = \frac{1}{2}\left(\operatorname{tr} B(x) \pm \sqrt{(\operatorname{tr} B(x))^2 - 4 \det B(x)}\right)$$

it is easy to see that both roots (or their real parts) have the same sign (sign of tr $B(x)$) if and only if det $B(x) > 0$.

To calculate an approximation to the one-dimensional slow invariant manifold

$$y_1 = \bar{h}(x, \varepsilon), \quad y_2 = \bar{\bar{h}}(x, \varepsilon)$$

from the invariance equations

$$\varepsilon \frac{\partial \bar{h}(x, \varepsilon)}{\partial x} f(x, \bar{h}(x, \varepsilon), \bar{\bar{h}}(x, \varepsilon), \varepsilon) = g_1(x, \bar{h}(x, \varepsilon), \bar{\bar{h}}(x, \varepsilon), \varepsilon),$$

$$\varepsilon \frac{\partial \bar{\bar{h}}(x, \varepsilon)}{\partial x} f(x, \bar{h}(x, \varepsilon), \bar{\bar{h}}(x, \varepsilon), \varepsilon) = g_2(x, \bar{h}(x, \varepsilon), \bar{\bar{h}}(x, \varepsilon), \varepsilon),$$

we substitute the formal expansions

$$\bar{h}(x, \varepsilon) = \bar{\phi}(x) + \varepsilon \bar{h}_1(x) + O(\varepsilon^2)$$

and

$$\bar{\bar{h}}(x, \varepsilon) = \bar{\bar{\phi}}(x) + \varepsilon \bar{\bar{h}}_1(x) + O(\varepsilon^2)$$

into these equations to obtain

$$\varepsilon \frac{d\bar{\phi}(x)}{dx} f(x, \bar{\phi}(x), \bar{\bar{\phi}}(x), 0) = \varepsilon b_{11}(x)\bar{h}_1(x) + \varepsilon b_{12}(x)\bar{\bar{h}}_1(x)$$

$$+ \varepsilon \frac{\partial g_1}{\partial \varepsilon}(x, \bar{\phi}(x), \bar{\bar{\phi}}(x), 0) + O(\varepsilon^2), \qquad (3.46)$$

$$\varepsilon \frac{d\bar{\bar{\phi}}(x)}{dx} f(x, \bar{\phi}(x), \bar{\bar{\phi}}(x), 0) = \varepsilon b_{21}(x)\bar{h}_1(x) + \varepsilon b_{22}(x)\bar{\bar{h}}_1(x)$$

$$+ \varepsilon \frac{\partial g_2}{\partial \varepsilon}(x, \bar{\phi}(x), \bar{\bar{\phi}}(x), 0) + O(\varepsilon^2).$$

It is a straightforward calculation now to obtain the following expressions

$$\bar{h}_1(x) = \frac{a_1(x)b_{22}(x) - a_2(x)b_{12}(x)}{\det B(x)},$$

$$\bar{\bar{h}}_1(x) = \frac{a_2(x)b_{11}(x) - a_1(x)b_{21}(x)}{\det B(x)},$$

where

$$a_1(x) = \frac{d\bar{\phi}(x)}{dx} f(x, \bar{\phi}(x), \bar{\bar{\phi}}(x), 0) - \frac{\partial g_1}{\partial \varepsilon}(x, \bar{\phi}(x), \bar{\bar{\phi}}(x), 0),$$

and

$$a_2(x) = \frac{d\bar{\bar{\phi}}(x)}{dx} f(x, \bar{\phi}(x), \bar{\bar{\phi}}(x), 0) - \frac{\partial g_2}{\partial \varepsilon}(x, \bar{\phi}(x), \bar{\bar{\phi}}(x), 0).$$

Thus, we construct the slow invariant manifold

$$y_1 = \bar{h}(x, \varepsilon) = \bar{\phi}(x) + \varepsilon \bar{h}_1(x) + O(\varepsilon^2),$$

$$y_2 = \bar{\bar{h}}(x, \varepsilon) = \bar{\bar{\phi}}(x) + \varepsilon \bar{\bar{h}}_1(x) + O(\varepsilon^2),$$

and the motion of system (3.45) on the slow invariant manifold is described by

$$\frac{dx}{dt} = f(x, \bar{h}(x, \varepsilon), \bar{\bar{h}}(x, \varepsilon), \varepsilon).$$

3.4.2 Cooperative Phenomenon

Consider now an example of a $1 + 2$ system: the so called *cooperative phenomenon* [119]. J.D. Murray [119] describes the situation as follows. A model consists of an enzyme molecule E which binds a substrate molecule S to form a single bound substrate-enzyme complex C_1. This complex C_1 not only breaks down to form a product P and enzyme E again, it also combines with another substrate molecule to form a dual bound substrate-enzyme complex C_2. This C_2 complex breaks down to form a product P and the single bound complex C_1. A reaction mechanism for this model is

$$S + E \underset{k_{-1}}{\overset{k_1}{\rightleftharpoons}} C_1 \overset{k_2}{\rightarrow} E + P, \tag{3.47}$$

$$S + C_1 \underset{k_{-3}}{\overset{k_3}{\rightleftharpoons}} C_2 \overset{k_4}{\rightarrow} C_1 + P, \tag{3.48}$$

where k's are the rate constants as indicated.

With lower case letters denoting concentrations, the mass action law applied to (3.47), (3.48) gives the differential equations in dimensional form

$$\frac{ds}{dt} = -k_1\, s\, e + (k_{-1} - k_3 s)c_1 + k_{-3}c_2, \tag{3.49}$$

$$\frac{dc_1}{dt} = k_1\, s\, e - (k_{-1} + k_2 + k_3 s)c_1 + (k_{-3} + k_4)c_2, \tag{3.50}$$

$$\frac{dc_2}{dt} = k_3\, s\, c_1 - (k_{-3} + k_4)c_2 \tag{3.51}$$

$$\frac{de}{dt} = -k_1\, s\, e + (k_{-1} + k_2)c_1, \tag{3.52}$$

$$\frac{dp}{dt} = k_2\, c_1 + k_4\, c_2. \tag{3.53}$$

Appropriate initial conditions are

$$s(0) = s_0, \quad e(0) = e_0, \quad c_1(0) = c_2(0) = p(0) = 0, \tag{3.54}$$

i.e., the initial concentrations of the substrate S and enzyme E are specified to be non-zero. The conservation of the enzyme is obtained by adding the 2nd, 3rd, and 4th equations in (3.49)–(3.53) and using the initial conditions; it is

$$\frac{d}{dt}(c_1 + c_2 + e) = 0 \;\Rightarrow\; c_1 + c_2 + e = e_0. \tag{3.55}$$

Equation (3.53) for the product $p(t)$ is uncoupled and given by integration, once c_1 and c_2 have been found. Thus, by using (3.55), the resulting system is

$$\frac{ds}{dt} = -k_1\, s\, e_0 + (k_{-1} + k_1 s - k_3 s)c_1 + (k_1 s + k_{-3})c_2, \tag{3.56}$$

$$\frac{dc_1}{dt} = k_1\, s\, e_0 - (k_{-1} + k_2 + k_1 s + k_3 s)c_1 + (k_{-3} + k_4 - k_1 s)c_2, \tag{3.57}$$

$$\frac{dc_2}{dt} = k_3\, s\, c_1 - (k_{-3} + k_4)c_2. \tag{3.58}$$

We nondimensionalize the system by introducing the dimensionless variables

$$x(\tau) = \frac{s(t)}{s_0}, \quad y_1(\tau) = \frac{c_1(t)}{e_0}, \quad y_2(\tau) = \frac{c_2(t)}{e_0},$$

the dimensionless time

$$\tau = k_1\, e_0\, t,$$

and the dimensionless parameters

$$\varepsilon = \frac{e_0}{s_0}, \quad a_1 = \frac{k_{-1}}{k_1 s_0}, \quad a_2 = \frac{k_2}{k_1 s_0}, \quad a_3 = \frac{k_3}{k_1}, \quad a_4 = \frac{k_{-3}}{k_1 s_0}, \quad a_5 = \frac{k_4}{k_1 s_0}.$$

Then (3.56)–(3.58) become

$$\frac{dx}{d\tau} = -x + (x - a_3 x + a_1)y_1 + (a_4 + x)y_2 = f(x, y_1, y_2), \tag{3.59}$$

$$\varepsilon \frac{dy_1}{d\tau} = x - (x + a_3 x + a_1 + a_2)y_1 + (a_4 + a_5 - x)y_2 = g_1(x, y_1, y_2), \tag{3.60}$$

$$\varepsilon \frac{dy_2}{d\tau} = a_3 x y_1 - (a_4 + a_5)y_2 = g_2(x, y_1, y_2), \tag{3.61}$$

with initial conditions

$$x(0) = 1, \quad y_1(0) = y_2(0) = 0. \tag{3.62}$$

This problem, as with the Michaelis–Menten problem, is singularly perturbed for $0 < \varepsilon \ll 1$. Note, that the origin $x = y_1 = y_2 = 0$ is the unique equilibrium of (3.59)–(3.61) with nonnegative x, y_1 and y_2. Another equilibrium has the coordinates

$$x = x_s = -\frac{a_2(a_4 + a_5)}{a_3 a_5}, \quad y_1 = \frac{x_s(a_4 + a_5)}{\delta}, \quad y_2 = \frac{x_s^2}{\delta(a_4 + a_5)},$$

where $\delta = (a_4 + a_5)(x_s + a_1 + a_2) + a_3 x_s^2$. Since $x_s < 0$ this equilibrium doesn't correspond to a physical situation.

Now we use the results of the previous section to calculate the approximation of the one-dimensional slow invariant manifold and the equation which describes the flow on this manifold.

The corresponding degenerate system is

$$\frac{dx}{dt} = -x + (x - a_3 x + a_1)y_1 + (a_4 + x)y_2,$$

$$0 = x - (x + a_3 x + a_1 + a_2)y_1 + (a_4 + a_5 - x)y_2,$$

$$0 = a_3 x y_1 - (a_4 + a_5)y_2.$$

The last two equations give the unique solution

$$y_1 = \bar{\phi}(x) = x/\Delta,$$

$$y_2 = \bar{\bar{\phi}}(x) = a_3 a x^2/\Delta.$$

Here Δ/a is the determinant of the Jacobian matrix

$$B = \begin{pmatrix} \frac{\partial g_1}{\partial y_1} & \frac{\partial g_1}{\partial y_2} \\ \frac{\partial g_2}{\partial y_1} & \frac{\partial g_2}{\partial y_2} \end{pmatrix} = \begin{pmatrix} -x - a_3 x - a_1 - a_2 & a_4 + a_5 - x \\ a_3 x & - a_4 - a_5 \end{pmatrix},$$

where $\Delta = x + a_1 + a_2 + a_3 a x^2$ and $a = (a_4 + a_5)^{-1}$. The slow curve is stable since the $-\operatorname{tr} B(x)$ and $\det B(x)$ are positive.

To calculate the approximation to the one-dimensional slow invariant manifold

$$y_1 = \bar{h}(x, \varepsilon) = \bar{\phi}(x) + \varepsilon \bar{h}_1(x) + O(\varepsilon^2),$$

$$y_2 = \bar{\bar{h}}(x, \varepsilon) = \bar{\bar{\phi}}(x) + \varepsilon \bar{\bar{h}}_1(x) + O(\varepsilon^2),$$

we rewrite the invariance equations (3.46) for the system (3.59)–(3.61):

$$\varepsilon \frac{d\bar{\phi}(x)}{dx}[-x + (x - a_3 x + a_1)(\bar{\phi}(x) + \varepsilon \bar{h}_1(x) + \varepsilon^2 \ldots)$$

$$+ (a_4 + x)(\bar{\bar{\phi}}(x) + \varepsilon \bar{\bar{h}}_1(x) + \varepsilon^2 \ldots)]$$

$$= x - (x + a_3 x + a_1 + a_2)(\bar{\phi}(x) + \varepsilon \bar{h}_1(x) + \varepsilon^2 \ldots)$$

$$+ (a_4 + a_5 - x)(\bar{\bar{\phi}}(x) + \varepsilon \bar{\bar{h}}_1(x) + \varepsilon^2 \ldots),$$

$$\varepsilon \frac{d\bar{\bar{\phi}}(x)}{dx}[-x + (x - a_3 x + a_1)(\bar{\phi}(x) + \varepsilon \bar{h}_1(x) + \varepsilon^2 \ldots)$$

$$+ (a_4 + x)(\bar{\bar{\phi}}(x) + \varepsilon \bar{\bar{h}}_1(x) + \varepsilon^2 \ldots)]$$

$$= a_3 x (\bar{\phi}(x) + \varepsilon \bar{h}_1(x) + \varepsilon^2 \ldots)$$

$$- (a_4 + a_5)(\bar{\bar{\phi}}(x) + \varepsilon \bar{\bar{h}}_1(x) + \varepsilon^2 \ldots).$$

Using the formulae

$$\frac{d\bar{\phi}(x)}{dx} = (a_1 + a_2 - a_3 a x^2)/\Delta^2,$$

$$\frac{d\bar{\bar{\phi}}(x)}{dx} = a a_3 x [2(a_1 + a_2) + x]/\Delta^2,$$

we solve for $\bar{h}_1(x), \bar{\bar{h}}_1(x)$ to get

$$\bar{h}_1(x) = -\frac{(a_2 + a_3 a_5 ax)ax}{\Delta^4}\left[a_3 ax^3 - (a_1 + a_2)(a_4 + a_5 + 2a_3 x - 2a_3 ax^2)\right],$$

$$\bar{\bar{h}}_1(x) = -\frac{(a_2 + a_3 a_5 ax)ax}{\Delta^4}\left[-a_3 ax^3 - (a_1 + a_2)a_3 x\left(1 + 2a(a_1 + a_2)\right.\right.$$

$$\left.\left. +(3 + 2a_3)ax\right)\right].$$

The flow on the stable slow invariant manifold

$$y_1 = \frac{x}{\Delta} - \varepsilon\frac{(a_2 + a_3 a_5 ax)ax}{\Delta^4}\left[a_3 ax^3 - (a_1 + a_2)(a_4 + a_5 + 2a_3 x - 2a_3 ax^2)\right]$$

$$+O(\varepsilon^2),$$

$$y_2 = \frac{a_3 ax^2}{\Delta} - \varepsilon\frac{(a_2 + a_3 a_5 ax)ax}{\Delta^4}$$

$$\times\left[-a_3 ax^3 - (a_1 + a_2)a_3 x\left(1 + 2a(a_1 + a_2) + (3 + 2a_3)ax\right)\right] + O(\varepsilon^2)$$

is given by

$$\frac{dx}{d\tau} = -x + (x - a_3 x + a_1)(\bar{\phi}(x) + \varepsilon\bar{h}_1(x)) + (a_4 + x)(\bar{\bar{\phi}}(x) + \varepsilon\bar{\bar{h}}_1(x))$$

$$+O(\varepsilon^2))$$

or

$$\frac{dx}{d\tau} = -\frac{x(a_2 + a_3 a_5 ax)}{\Delta} + \varepsilon\left[(x - a_3 x + a_1)\bar{h}_1(x) + (a_4 + x)\bar{\bar{h}}_1(x)\right] + O(\varepsilon^2).$$

This last equation implies that the origin is an asymptotically stable equilibrium because the coefficient $-\frac{a_2}{\Delta}$ is negative.

On Fig. 3.6 we can see that the trajectory approaches the slow invariant manifold very quickly and then follows along it to the origin, as $t \to \infty$.

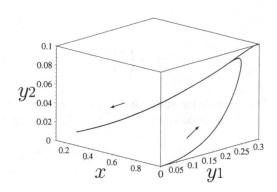

Fig. 3.6 The slow invariant manifold and the trajectory of (3.59)–(3.61) with initial conditions (3.62) for $a_1 = 1$, $a_2 = 1$, $a_3 = 1$, $a_4 = 1$, $a_5 = 2$, and $\varepsilon = 0.01$. The *arrows* indicate increasing time

If we return to dimensional variables, we obtain the generalization of the Michaelis–Menten law for the case of the two enzyme–substrate complexes, see [119].

3.4.3 Cooperative Phenomenon: Another Approach

We take Eqs. (3.59)–(3.61) and apply the approach developed in Sect. 2.5 to construct an approximation to the slow invariant manifold and the flow on it.

We refer back to Eq. (2.26) viz.,

$$\dot{x} = \zeta(x,t,\varepsilon) + F(x,t,\varepsilon)y,$$
$$\varepsilon\dot{y} = \xi(x,t,\varepsilon) + G(x,t,\varepsilon)y.$$

Then

$$\zeta = \zeta_0 = -x; \quad F = F_0 = \left(x - a_3x + a_1, a_4 + x\right);$$

$$\xi = \xi_0 = \begin{pmatrix} x \\ 0 \end{pmatrix}; \quad G = G_0 = \begin{pmatrix} -x - a_3x - a_1 - a_2 & a_4 + a_5 - x \\ a_3x & -a_4 - a_5 \end{pmatrix}.$$

Note that $\zeta_1 = 0$, $F_1 = 0$, $\xi_1 = 0$, and $G_1 = 0$.
The inverse matrix is

$$G_0^{-1} = \frac{a}{\Delta} \begin{pmatrix} -a_4 - a_5 & x - a_4 - a_5 \\ -a_3x & -x - a_3x - a_1 - a_2 \end{pmatrix},$$

where $a = (a_4 + a_5)^{-1}$, $\Delta = x + a_1 + a_2 + a_3ax^2$. We obtain the zero order approximation $\phi(x)$ to the slow invariant manifold

$$y = h(x,\varepsilon) = \begin{pmatrix} \bar{h}(x,\varepsilon) \\ \bar{\bar{h}}(x,\varepsilon) \end{pmatrix} = \begin{pmatrix} \bar{\phi}(x) + \varepsilon\bar{h}_1(x) + O(\varepsilon^2) \\ \bar{\bar{\phi}}(x) + \varepsilon\bar{\bar{h}}_1(x) + O(\varepsilon^2) \end{pmatrix}$$

viz.,

$$\phi = \begin{pmatrix} \bar{\phi}(x) \\ \bar{\bar{\phi}}(x) \end{pmatrix} = -G_0^{-1}\xi_0 = \frac{1}{\Delta} \begin{pmatrix} x \\ a_3ax^2 \end{pmatrix}.$$

The zero order approximation to the flow on this slow invariant manifold is described by the equation $\dot{x} = \zeta_0 + F_0\phi$, where

$$\zeta_0 + F_0\phi = -\frac{x(a_2 + a_3a_5ax)}{\Delta}.$$

Taking into account

$$\frac{d\phi}{dx} = \frac{1}{\Delta^2} \left(\begin{array}{c} a_1 + a_2 - a_3 a x^2 \\ a a_3 (a_1 + a_2) 2x + a a_3 x^2 \end{array} \right),$$

we obtain the first order correction to the slow invariant manifold

$$h_1 = \left(\begin{array}{c} \bar{h}_1(x) \\ \bar{\bar{h}}_1(x) \end{array} \right) = G_0^{-1} \frac{d\phi}{dx} (\zeta_0 + F_0 \phi)$$

$$= -\frac{(a_2 + a_3 a_5 a x) a x}{\Delta^4} \left(\begin{array}{c} a_3 a x^3 - (a_1 + a_2)(a_4 + a_5 + 2a_3 x - 2a_3 a x^2) \\ -a_3 a x^3 - (a_1 + a_2) a_3 x \left(1 + 2a(a_1 + a_2) + (3 + 2a_3) a x\right) \end{array} \right)$$

and the first order correction to r.h.s. of the equation for the flow on this manifold viz.,

$$F_0 h_1 = (x - a_3 x + a_1) \bar{h}_1(x) + (a_4 + x) \bar{\bar{h}}_1(x).$$

We have thus obtained the same representation for the first order approximation of the slow invariant manifold and the flow on it.

3.5 2+2

3.5.1 Theoretical Background

We now consider the autonomous differential system with two slow variables x_1, x_2 and two fast variables y_1, y_2

$$\frac{dx_1}{dt} = f_1(x_1, x_2, y_1, y_2),$$

$$\frac{dx_2}{dt} = f_2(x_1, x_2, y_1, y_2),$$

$$\varepsilon \frac{dy_1}{dt} = g_1(x_1, x_2, y_1, y_2),$$

$$\varepsilon \frac{dy_2}{dt} = g_2(x_1, x_2, y_1, y_2),$$

(3.63)

and with a small positive parameter ε. For simplicity we consider the case when the functions on the r.h.s. of (3.63) do not depend on ε. The corresponding degenerate system is

$$\frac{dx_1}{dt} = f_1(x_1, x_2, y_1, y_2),$$

$$\frac{dx_2}{dt} = f_2(x_1, x_2, y_1, y_2),$$

$$0 = g_1(x_1, x_2, y_1, y_2),$$

$$0 = g_2(x_1, x_2, y_1, y_2).$$

The last two equations describe a two-dimensional slow surface. Suppose that these equations are solved for y_1, y_2, i.e.,

$$y_1 = \bar{\phi}(x_1, x_2), \quad y_2 = \bar{\bar{\phi}}(x_1, x_2).$$

As previously, we consider the Jacobian matrix along the slow surface, i.e.,

$$B = B(x_1, x_2) = \left. \begin{pmatrix} \dfrac{\partial g_1}{\partial y_1} & \dfrac{\partial g_1}{\partial y_2} \\ \dfrac{\partial g_2}{\partial y_1} & \dfrac{\partial g_2}{\partial y_2} \end{pmatrix} \right|_{y_1 = \bar{\phi}(x_1, x_2), \quad y_2 = \bar{\bar{\phi}}(x_1, x_2)}.$$

If both two roots of the characteristic polynomial

$$\det(B(x_1, x_2) - \lambda \mathbf{I}) = 0,$$

where $\mathbf{I}$ is the identity matrix, have negative real parts, then the slow surface is stable. Let

$$b_{ij}(x_1, x_2) = \left. \frac{\partial g_i}{\partial y_j} \right|_{y_1 = \bar{\phi}(x_1, x_2), \quad y_2 = \bar{\bar{\phi}}(x_1, x_2)}, \quad i, j = 1, 2,$$

$$\operatorname{tr} B(x_1, x_2) = b_{11}(x_1, x_2) + b_{22}(x_1, x_2),$$

$$\det B(x_1, x_2) = b_{11}(x_1, x_2) b_{22}(x_1, x_2) - b_{12}(x_1, x_2) b_{21}(x_1, x_2),$$

where $\operatorname{tr} B(x_1, x_2)$ is the trace and $\det B(x_1, x_2)$ is the determinant of $B(x_1, x_2)$. As before the condition for stability of the slow surface is the positivity of the coefficients of the quadratic characteristic polynomial, which are $-\operatorname{tr} B(x_1, x_2)$ and $\det B(x_1, x_2)$.

To calculate the approximate two-dimensional slow invariant manifold

$$y_1 = \bar{h}(x_1, x_2, \varepsilon), \quad y_2 = \bar{\bar{h}}(x_1, x_2, \varepsilon)$$

from the invariance equations

$$\varepsilon \frac{\partial \bar{h}(x_1, x_2, \varepsilon)}{\partial x_1} f_1(x_1, x_2, \bar{h}(x_1, x_2, \varepsilon), \bar{\bar{h}}(x_1, x_2, \varepsilon))$$

$$+ \varepsilon \frac{\partial \bar{h}(x_1, x_2, \varepsilon)}{\partial x_2} f_2(x_1, x_2, \bar{h}(x_1, x_2, \varepsilon), \bar{\bar{h}}(x_1, x_2, \varepsilon))$$

$$= g_1(x_1, x_2, \bar{h}(x_1, x_2, \varepsilon), \bar{\bar{h}}(x_1, x_2, \varepsilon)), \tag{3.64}$$

and

$$\varepsilon \frac{\partial \bar{\bar{h}}(x_1, x_2, \varepsilon)}{\partial x_1} f_1(x_1, x_2, \bar{h}(x_1, x_2, \varepsilon), \bar{\bar{h}}(x_1, x_2, \varepsilon))$$

$$+ \varepsilon \frac{\partial \bar{\bar{h}}(x_1, x_2, \varepsilon)}{\partial x_2} f_2(x_1, x_2, \bar{h}(x_1, x_2, \varepsilon), \bar{\bar{h}}(x_1, x_2, \varepsilon))$$

$$= g_2(x_1, x_2, \bar{h}(x_1, x_2, \varepsilon), \bar{\bar{h}}(x_1, x_2, \varepsilon)), \tag{3.65}$$

we substitute the formal expansions

$$\bar{h}(x_1, x_2, \varepsilon) = \bar{\phi}(x_1, x_2) + \varepsilon \bar{h}_1(x_1, x_2) + O(\varepsilon^2),$$

$$\bar{\bar{h}}(x_1, x_2, \varepsilon) = \bar{\bar{\phi}}(x_1, x_2) + \varepsilon \bar{\bar{h}}_1(x_1, x_2) + O(\varepsilon^2)$$

into these equations. The result is

$$\varepsilon \frac{\partial \bar{\phi}(x_1, x_2)}{\partial x_1} f_1(x_1, x_2, \bar{\phi}(x_1, x_2), \bar{\bar{\phi}}(x_1, x_2))$$

$$+ \varepsilon \frac{\partial \bar{\phi}(x_1, x_2)}{\partial x_2} f_2(x_1, x_2, \bar{\phi}(x_1, x_2), \bar{\bar{\phi}}(x_1, x_2)) \tag{3.66}$$

$$= \varepsilon b_{11}(x_1, x_2) \bar{h}_1(x_1, x_2) + \varepsilon b_{12}(x_1, x_2) \bar{\bar{h}}_1(x_1, x_2) + O(\varepsilon^2),$$

and

$$\varepsilon \frac{\partial \bar{\bar{\phi}}(x_1, x_2)}{\partial x_1} f_1(x_1, x_2, \bar{\phi}(x_1, x_2), \bar{\bar{\phi}}(x_1, x_2))$$

$$+ \varepsilon \frac{\partial \bar{\bar{\phi}}(x_1, x_2)}{\partial x_2} f_2(x_1, x_2, \bar{\phi}(x_1, x_2), \bar{\bar{\phi}}(x_1, x_2)) \tag{3.67}$$

$$= \varepsilon b_{21}(x_1, x_2) \bar{h}_1(x_1, x_2) + \varepsilon b_{22}(x_1, x_2) \bar{\bar{h}}_1(x_1, x_2) + O(\varepsilon^2).$$

It is a straightforward exercise now to obtain the expressions

$$\bar{h}_1(x_1, x_2) = \frac{a_1(x_1, x_2)b_{22}(x_1, x_2) - a_2(x_1, x_2)b_{12}(x_1, x_2)}{B(x_1, x_2)},$$

$$\bar{\bar{h}}_1(x_1, x_2) = \frac{a_2(x_1, x_2)b_{11}(x_1, x_2) - a_1(x_1, x_2)b_{21}(x_1, x_2)}{B(x_1, x_2)},$$

where

$$a_1(x_1, x_2) = \frac{\partial\bar{\phi}(x_1, x_2)}{\partial x_1} f_1(x_1, x_2, \bar{\phi}(x_1, x_2), \bar{\bar{\phi}}(x_1, x_2))$$

$$+ \frac{\partial\bar{\phi}(x_1, x_2)}{\partial x_2} f_2(x_1, x_2, \bar{\phi}(x_1, x_2), \bar{\bar{\phi}}(x_1, x_2)),$$

and

$$a_2(x_1, x_2) = \frac{\partial\bar{\bar{\phi}}(x_1, x_2)}{\partial x_1} f_1(x_1, x_2, \bar{\phi}(x_1, x_2), \bar{\bar{\phi}}(x_1, x_2))$$

$$+ \frac{\partial\bar{\bar{\phi}}(x_1, x_2)}{\partial x_2} f_2(x_1, x_2, \bar{\phi}(x_1, x_2), \bar{\bar{\phi}}(x_1, x_2)).$$

We have calculated $\bar{\phi}(x_1, x_2)$, $\bar{h}_1(x_1, x_2)$, $\bar{\bar{\phi}}(x_1, x_2)$ and $\bar{\bar{h}}_1(x_1, x_2)$ and thus have an approximation to the two-dimensional slow invariant manifold. The flow on this manifold is now calculated from the first two equations in (3.63), taking into account

$$y_1 = \bar{h}(x_1, x_2, \varepsilon) = \bar{\phi}(x_1, x_2) + \varepsilon\bar{h}_1(x_1, x_2) + O(\varepsilon^2)$$

and

$$y_2 = \bar{\bar{h}}(x_1, x_2, \varepsilon) = \bar{\bar{\phi}}(x_1, x_2) + \varepsilon\bar{\bar{h}}_1(x_1, x_2) + O(\varepsilon^2).$$

3.5.2 Enzyme–Substrate-Inhibitor System

In this section a enzyme–substrate reaction [118] is considered as an example of a $2 + 2$ system. The reaction consists of an enzyme E with a single reaction site (many enzymes have several such sites) for which two substrates compete and form one of two complexes. These break down to give two products and the original enzyme. When one substrate combines with the enzyme it means, in effect, that it is inhibiting the other substrate's reaction with that enzyme. The reactions can be written schematically as

$$S + E \underset{k_{-1}}{\overset{k_1}{\rightleftharpoons}} ES \overset{k_2}{\rightarrow} P_S + E, \tag{3.68}$$

$$I + E \underset{k_{-3}}{\overset{k_3}{\rightleftharpoons}} EI \overset{k_4}{\rightarrow} P_I + E, \tag{3.69}$$

where S and I are the two substrates, which compete for the same enzyme E, and P_S and P_I are the products of two enzyme–substrate reactions.

When two substrates are competing for the same enzyme site, the reaction system (3.68) and (3.69) is said to be *fully competitive*. In such reactions one or other of the substrates can be singled out for its reaction rate e.g., $r_0 = \left.\dfrac{ds}{dt}\right|_{t=0}$ to be measured by an experiment (see more details in [118]). The one so singled out is called the *substrate* and the other the *inhibitor*. We choose the inhibitor to be I and its reaction to be (3.69).

Applying the law of mass action to (3.68), (3.69) gives the kinetic equations for the concentrations of the reactants. Since we shall be interested primarily in the rates of the reactions of S and I, we do not need the equations for the products; only the rate constants k_2 and k_4 in (3.68), (3.69) are involved. Thus we need only consider the kinetic equations for the substrate, inhibitor, and enzyme complex whose concentrations as functions of time t are denoted by

$$s(t) = [S], \; i(t) = [I], \; e(t) = [E],$$
$$c_s(t) = [ES], \; c_i(t) = [EI]. \tag{3.70}$$

The kinetic equations for the concentrations for the reactions (3.68), (3.69), see [118], are

$$\frac{ds}{dt} = -k_1 s e + k_{-1} c_s, \tag{3.71}$$

$$\frac{dc_s}{dt} = k_1 s e - (k_{-1} + k_2)c_s, \tag{3.72}$$

$$\frac{di}{dt} = -k_3 i e + k_{-3} c_i, \tag{3.73}$$

$$\frac{dc_i}{dt} = k_3 i e - (k_{-3} + k_4)c_i \tag{3.74}$$

$$\frac{de}{dt} = -k_1 s e - k_3 i e + (k_{-1} + k_2)c_s + (k_{-3} + k_4)c_i. \tag{3.75}$$

Appropriate initial conditions for Eqs. (3.71)–(3.75) are that there are no enzyme complexes initially but s, i, and e are prescribed, that is

$$s(0) = s_0, \; i(0) = i_0, \; e(0) = e_0, \; c_s(0) = c_i(0) = 0. \tag{3.76}$$

The conservation equation for the enzyme e is obtained immediately by adding (3.72), (3.74), (3.75) and using the initial conditions (3.76) to get

$$\frac{d}{dt}(c_s + c_i + e) = 0 \Rightarrow c_s + c_i + e = e_0. \tag{3.77}$$

Eliminating e from (3.71)–(3.75) by using (3.77) gives four equations for s, i, c_s and c_i. We now introduce nondimensional variables and parameters by

$$x_1(\tau) = \frac{s(t)}{s_0}, \quad x_2(\tau) = \frac{i(t)}{i_0}, \quad y_1(\tau) = \frac{c_s(t)}{e_0}, \quad y_2(\tau) = \frac{c_i(t)}{e_0},$$

$$\tau = k_1 e_0 t, \quad \varepsilon = \frac{e_0}{s_0}, \quad \beta = \frac{i_0}{s_0}, \quad \gamma = \frac{k_3}{k_1}, \tag{3.78}$$

$$K_s = \frac{k_{-1} + k_2}{k_1 s_0}, \quad K_i = \frac{k_{-3} + k_4}{k_3 i_0}, \quad L_s = \frac{k_2}{k_1 s_0}, \quad L_i = \frac{k_4}{k_3 i_0}.$$

Then the four equations for s, i, c_s and c_i become the four dimensionless equations

$$\frac{dx_1}{d\tau} = -x_1 + (x_1 + K_s - L_s)y_1 + x_1 y_2 = f_1(x_1, x_2, y_1, y_2), \tag{3.79}$$

$$\frac{dx_2}{d\tau} = \gamma[-x_2 + x_2 y_1 + (x_2 + K_i - L_i)y_2] = f_2(x_1, x_2, y_1, y_2), \tag{3.80}$$

$$\varepsilon \frac{dy_1}{d\tau} = x_1 - (x_1 + K_s)y_1 - x_1 y_2 = g_1(x_1, x_2, y_1, y_2), \tag{3.81}$$

$$\varepsilon \frac{dy_2}{d\tau} = \beta\gamma[x_2 - x_2 y_1 - (x_2 + K_i)y_2] = g_2(x_1, x_2, y_1, y_2), \tag{3.82}$$

with initial conditions

$$x_1(0) = x_2(0) = 1, y_1(0) = y_2(0) = 0. \tag{3.83}$$

We use the results of the previous section to calculate the approximate two-dimensional slow invariant manifold and the equation that describes the flow on this manifold with the assumption that $0 < \varepsilon \ll 1$.

The degenerate system is

$$\frac{dx_1}{d\tau} = -x_1 + (x_1 + K_s - L_s)y_1 + x_1 y_2, \tag{3.84}$$

$$\frac{dx_2}{d\tau} = \gamma[-x_2 + x_2 y_1 + (x_2 + K_i - L_i)y_2], \tag{3.85}$$

$$0 = x_1 - (x_1 + K_s)y_1 - x_1 y_2, \tag{3.86}$$

$$0 = \beta\gamma[x_2 - x_2 y_1 - (x_2 + K_i)y_2], \tag{3.87}$$

The last two equations give the unique solution

$$y_1 = \bar{\phi}(x_1, x_2) = K_i x_1 / \Delta,$$

$$y_2 = \bar{\bar{\phi}}(x_1, x_2) = K_s x_2 / \Delta.$$

Here $\Delta\beta\gamma$ is the determinant of the Jacobian matrix

$$B(x_1, x_2) = \begin{pmatrix} \dfrac{\partial g_1}{\partial y_1} & \dfrac{\partial g_1}{\partial y_2} \\ \dfrac{\partial g_2}{\partial y_1} & \dfrac{\partial g_2}{\partial y_2} \end{pmatrix} = \begin{pmatrix} -x_1 - K_s & -x_1 \\ -\beta\gamma x_2 & -\beta\gamma(x_2 + K_i) \end{pmatrix},$$

where $\Delta = K_s x_2 + K_i x_1 + K_s K_i$. The slow surface is stable since the $-\mathrm{tr}\, B(x_1, x_2)$ and $\det B(x_1, x_2)$ are positive.

To calculate the approximations to the two-dimensional slow invariant manifold, we assume

$$y_1 = \bar{h}(x_1, x_2, \varepsilon) = \bar{\phi}(x_1, x_2) + \varepsilon \bar{h}_1(x_1, x_2) + O(\varepsilon^2),$$

and

$$y_2 = \bar{\bar{h}}(x_1, x_2, \varepsilon) = \bar{\bar{\phi}}(x_1, x_2) + \varepsilon \bar{\bar{h}}_1(x_1, x_2) + O(\varepsilon^2).$$

The invariance equations (3.67), (3.68) for the system (3.79)–(3.82) yield

$$\varepsilon \frac{\partial \bar{\phi}(x_1, x_2)}{\partial x_1}(-x_1 + (x_1 + K_s - L_s)(\bar{\phi}(x_1, x_2) + \varepsilon \bar{h}_1(x_1, x_2) + O(\varepsilon^2))$$

$$+ x_1(\bar{\bar{\phi}}(x_1, x_2) + \varepsilon \bar{\bar{h}}_1(x_1, x_2) + O(\varepsilon^2)))$$

$$+ \varepsilon \frac{\partial \bar{\phi}(x_1, x_2)}{\partial x_2}(\gamma[-x_2 + x_2(\bar{\phi}(x_1, x_2) + \varepsilon \bar{h}_1(x_1, x_2) + O(\varepsilon^2))$$

$$+ (x_2 + K_i - L_i)(\bar{\bar{\phi}}(x_1, x_2) + \varepsilon \bar{\bar{h}}_1(x_1, x_2) + O(\varepsilon^2))])$$

$$= x_1 - (x_1 + K_s)(\bar{\phi}(x_1, x_2) + \varepsilon \bar{h}_1(x_1, x_2) + O(\varepsilon^2))$$

$$- x_1(\bar{\bar{\phi}}(x_1, x_2) + \varepsilon \bar{\bar{h}}_1(x_1, x_2) + O(\varepsilon^2)),$$

and

$$\varepsilon \frac{\partial \bar{\bar{\phi}}(x_1, x_2)}{\partial x_1}(-x_1 + (x_1 + K_s - L_s)(\bar{\phi}(x_1, x_2) + \varepsilon \bar{h}_1(x_1, x_2) + O(\varepsilon^2))$$

$$+ x_1(\bar{\bar{\phi}}(x_1, x_2) + \varepsilon \bar{\bar{h}}_1(x_1, x_2) + O(\varepsilon^2)))$$

$$+\varepsilon\frac{\partial\bar{\bar{\phi}}(x_1,x_2)}{\partial x_2}(\gamma[-x_2 + x_2(\bar{\phi}(x_1,x_2) + \varepsilon\bar{h}_1(x_1,x_2) + O(\varepsilon^2))$$

$$+(x_2 + K_i - L_i)(\bar{\phi}(x_1,x_2) + \varepsilon\bar{\bar{h}}_1(x_1,x_2) + O(\varepsilon^2))])$$

$$= \beta\gamma[x_2 - x_2(\bar{\phi}(x_1,x_2) + \varepsilon\bar{h}_1(x_1,x_2) + O(\varepsilon^2))$$

$$-(x_2 + K_i)(\bar{\bar{\phi}}(x_1,x_2) + \varepsilon\bar{\bar{h}}_1(x_1,x_2) + O(\varepsilon^2))].$$

Using the formulae

$$\frac{\partial\bar{\phi}(x_1,x_2)}{\partial x_1} = K_i K_s(x_2 + K_i)/\Delta^2,$$

$$\frac{\partial\bar{\phi}(x_1,x_2)}{\partial x_2} = K_i K_s(-x_1)/\Delta^2,$$

$$\frac{\partial\bar{\bar{\phi}}(x_1,x_2)}{\partial x_1} = K_i K_s(-x_2)/\Delta^2,$$

$$\frac{\partial\bar{\bar{\phi}}(x_1,x_2)}{\partial x_2} = K_i K_s(x_1 + K_s)/\Delta^2,$$

we find the expressions for $\bar{h}_1(x), \bar{\bar{h}}_1(x)$:

$$\bar{h}_1(x) = \frac{K_i K_s}{\beta\gamma\Delta^4}(\beta\gamma(x_2 + K_i)Px_1 - x_1x_2Q),$$

$$\bar{\bar{h}}_1(x) = \frac{K_i K_s}{\beta\gamma\Delta^4}(-\beta\gamma x_1x_2P + (x_1 + K_s)x_2Q),$$

where

$$P = (K_i L_s - \gamma K_s L_i)x_2 + K_i^2 L_s,$$

$$Q = -(K_i L_s - \gamma K_s L_i)x_1 + \gamma K_s^2 L_i.$$

Consequently, the first order approximation to the flow on the slow invariant manifold is

$$\frac{dx_1}{d\tau} = \frac{K_i}{\Delta}\left[-L_sx_1 + \frac{\varepsilon K_s}{\beta\gamma\Delta^3}(\beta\gamma[K_ix_1 + (K_s - L_s)(x_2 + K_i)]Px_1\right.$$

$$\left. +L_sQx_1x_2)\right] + O(\varepsilon^2),$$

$$\frac{dx_2}{d\tau} = \frac{\gamma K_s}{\Delta}\left[-L_i x_2 + \frac{\varepsilon K_i}{\beta\gamma\Delta^3}(\beta\gamma x_1 x_2 L_i P \right.$$

$$\left. + [K_s x_2 + (K_i - L_i)(x_1 + K_s)]Q x_2) \right] + O(\varepsilon^2),$$

where the manifold is given by

$$y_1 = \frac{K_i x_1}{\Delta} + \varepsilon \frac{K_i K_s}{\beta\gamma\Delta^4}[\beta\gamma(x_2 + K_i)P x_1 - x_1 x_2 Q] + O(\varepsilon^2),$$

$$y_2 = \frac{K_s x_2}{\Delta} + \varepsilon \frac{K_i K_s}{\beta\gamma\Delta^4}[-\beta\gamma x_1 x_2 P + (x_1 + K_s)x_2 Q] + O(\varepsilon^2).$$

The first and second differential equations above for x_1 and x_2 imply that the origin

$$\begin{cases} x_1 = x_2 = 0, \\ y_1 = y_2 = 0 \end{cases}$$

is an asymptotically stable equilibrium because the coefficients $-\frac{K_i L_s}{\Delta_0}$ and $-\gamma\frac{K_s L_i}{\Delta_0}$ are negative. Here

$$\Delta_0 = \Delta|_{x_1=0, x_2=0} = K_s K_i \neq 0.$$

Figures 3.7, 3.8, 3.9, and 3.10 show that the trajectory of (3.79)–(3.83) approaches the slow invariant manifold very quickly and then follows along it as $\tau \to \infty$.

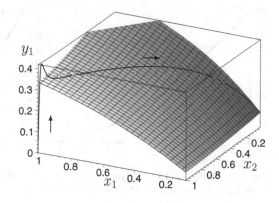

Fig. 3.7 The projections of the trajectory and slow invariant manifold of (3.79)–(3.83) on the plane $x_1 x_2 y_1$ for $K_i = K_s = 1$, $L_i = L_s = 1$, $\gamma = 2$, $\beta = 0.1$, and $\varepsilon = 0.01$

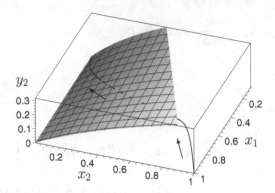

Fig. 3.8 The projections of the trajectory and slow invariant manifold of (3.79)–(3.83) on the plane $x_1 x_2 y_2$ for $K_i = K_s = 1$, $L_i = L_s = 1$, $\gamma = 2$, $\beta = 0.1$, and $\varepsilon = 0.01$

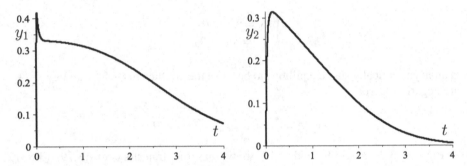

Fig. 3.9 The y_1- and y_2-components of the exact solution of (3.79)–(3.82) for $K_i = K_s = 1$, $L_i = L_s = 1$, $\gamma = 2$, $\beta = 0.1$, and $\varepsilon = 0.01$, and initial conditions $y_1(0) = y_2(0) = 0$

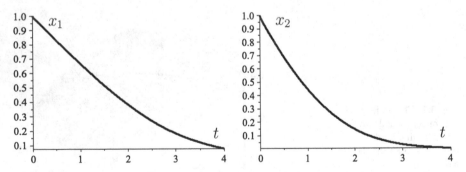

Fig. 3.10 The x_1- and x_2-components of the exact solution of (3.79)–(3.82) for $K_i = K_s = 1$, $L_i = L_s = 1$, $\gamma = 2$, $\beta = 0.1$, and $\varepsilon = 0.01$, and the initial conditions $x_1(0) = x_2(0) = 1$

3.5.3 Enzyme–Substrate-Exhibitor: The Another Approach

We consider again the model from Sect. 3.5.2 viz., Eqs. (3.79)–(3.82) and apply the technique from Sect. 2.5 to construct an approximation to the slow invariant manifold and the flow on it.

In this case we have

$$\zeta = \zeta_0 = \begin{pmatrix} -x_1 \\ -\gamma x_2 \end{pmatrix}; \quad F = F_0 = \begin{pmatrix} x_1 + K_s - L_s & x_1 \\ \gamma x_2 & \gamma(x_2 + K_i - L_i) \end{pmatrix};$$

$$\xi = \xi_0 = \begin{pmatrix} x_1 \\ \beta\gamma x_2 \end{pmatrix}; \quad G = G_0 = \begin{pmatrix} -x_1 - K_s & -x_1 \\ -\beta\gamma x_2 & -\beta\gamma(x_2 + K_i) \end{pmatrix}.$$

The inverse matrix is

$$G_0^{-1} = \frac{1}{\beta\gamma\Delta} \begin{pmatrix} -\beta\gamma(x_2 + K_i) & x_1 \\ \beta\gamma x_2 & -x_1 - K_s \end{pmatrix},$$

where

$$\Delta = K_s x_2 + K_i x_1 + K_s K_i.$$

We obtain the zero order approximation $\phi(x_1, x_2)$ to the slow invariant manifold

$$y = h(x_1, x_2, \varepsilon) = \begin{pmatrix} \bar{h}(x_1, x_2, \varepsilon) \\ \bar{\bar{h}}(x_1, x_2, \varepsilon) \end{pmatrix} = \begin{pmatrix} \bar{\phi}(x_1, x_2) + \varepsilon\bar{h}_1(x_1, x_2) + O(\varepsilon^2) \\ \bar{\bar{\phi}}(x_1, x_2) + \varepsilon\bar{\bar{h}}_1(x_1, x_2) + O(\varepsilon^2) \end{pmatrix};$$

$$\phi(x_1, x_2) = \begin{pmatrix} \bar{\phi}(x_1, x_2) \\ \bar{\bar{\phi}}(x_1, x_2) \end{pmatrix} = -G_0^{-1}\xi_0 = \frac{1}{\Delta} \begin{pmatrix} K_i x_1 \\ K_s x_2 \end{pmatrix}.$$

The zero order approximation to the flow on this slow invariant manifold is described by the equation $\dot{x} = \zeta_0 + F_0\phi$, where

$$\zeta_0 + F_0\phi = \begin{pmatrix} -x_1 \\ -\gamma x_2 \end{pmatrix}$$

$$+ \begin{pmatrix} x_1 + K_s - L_s & x_1 \\ \gamma x_2 & \gamma(x_2 + K_i - L_i) \end{pmatrix} \frac{1}{\Delta} \begin{pmatrix} K_i x_1 \\ K_s x_2 \end{pmatrix} = -\frac{1}{\Delta} \begin{pmatrix} K_i L_s x_1 \\ \gamma K_s L_i x_2 \end{pmatrix}.$$

Taking into account

$$\frac{\partial\phi}{\partial x} = \begin{pmatrix} \dfrac{\partial}{\partial x_1}\left(\dfrac{K_i x_1}{\Delta}\right) & \dfrac{\partial}{\partial x_2}\left(\dfrac{K_i x_1}{\Delta}\right) \\ \dfrac{\partial}{\partial x_1}\left(\dfrac{K_s x_2}{\Delta}\right) & \dfrac{\partial}{\partial x_2}\left(\dfrac{K_s x_2}{\Delta}\right) \end{pmatrix} = \frac{K_i K_s}{\Delta^2} \begin{pmatrix} x_2 + K_i & -x_1 \\ -x_2 & x_1 + K_s \end{pmatrix},$$

we obtain the first order correction to the slow invariant manifold

$$h_1 = \begin{pmatrix} \bar{h}_1 \\ \bar{\bar{h}}_1 \end{pmatrix} = G_0^{-1} \frac{\partial \phi}{\partial x} (\zeta_0 + F_0 \phi) = \frac{K_i K_s}{\beta \gamma \Delta^4} \begin{pmatrix} \beta \gamma (x_2 + K_i) P x_1 - x_1 x_2 Q \\ -\beta \gamma x_1 x_2 P + (x_1 + K_s) x_2 Q \end{pmatrix},$$

where

$$P = (K_i L_s - \gamma K_s L_i) x_2 + K_i^2 L_s,$$

and

$$Q = -(K_i L_s - \gamma K_s L_i) x_1 + \gamma K_s^2 L_i.$$

The first order approximation to the flow on the slow invariant manifold is described by the equation

$$\begin{pmatrix} \dot{x}_1 \\ \dot{x}_2 \end{pmatrix} = \zeta_0 + F_0 \phi + \varepsilon F_0 h_1 + O(\varepsilon^2),$$

where

$$F_0 h_1 = \frac{K_i K_s}{\beta \gamma \Delta^4} \begin{pmatrix} \beta \gamma [K_i x_1 + (K_s - L_s)(x_2 + K_i)] P x_1 + L_s Q x_1 x_2 \\ \beta \gamma^2 x_1 x_2 L_i P + \gamma [K_s x_2 + (K_i - L_i)(x_1 + K_s)] Q x_2 \end{pmatrix}.$$

Thus, we obtain the same representation for the first order approximation to the slow invariant manifold and the flow on it just as in Sect. 3.5.2.

Chapter 4
Representations of Slow Integral Manifolds

Abstract In constructing the asymptotic expansions of slow integral manifolds it is assumed that the degenerate equation ($\varepsilon = 0$) allows one to find the slow surface explicitly. In many problems this is not possible due to the fact that the degenerate equation is either a high degree polynomial or transcendental. In this situation many authors suggest the use of numerical methods. However, in many problems the slow surface can be described in parametric form, and then the slow integral manifold can be found in parametric form as asymptotic expansions. If this is not possible, it is necessary to use an implicit slow surface and obtain asymptotic representations for the slow integral manifold in an implicit form. Model examples, as well as examples borrowed from combustion theory, are treated.

4.1 Explicit and Implicit Slow Integral Manifolds

To describe the slow integral manifold for the system

$$\frac{dx}{dt} = f(x, y, t, \varepsilon),$$

$$\varepsilon \frac{dy}{dt} = g(x, y, t, \varepsilon),$$

(4.1)

we usually obtain the explicit representation $y = h(x, t, \varepsilon)$. In this case the approximation to $h(x, t, \varepsilon)$ may be obtained as an asymptotic expansion in powers of ε. However, it is generally not possible to find the function $y = \phi(x, t) = h(x, t, 0)$ exactly from the equation

$$g(x, y, t, 0) = 0.$$

However, in this case the slow integral manifold may be obtained in an implicit form

$$G(x, y, t, \varepsilon) = 0,$$

and the flow on this manifold is described by the differential equation

© Springer International Publishing Switzerland 2014

E. Shchepakina et al., *Singular Perturbations*, Lecture Notes in Mathematics 2114,
DOI 10.1007/978-3-319-09570-7_4

$$\frac{dx}{dt} = f(x, y, t, \varepsilon).$$

The invariance equation takes the form

$$G_y(x, y, t, \varepsilon) g(x, y, t, \varepsilon) + \varepsilon G_t(x, y, t, \varepsilon) + \varepsilon G_x(x, y, t, \varepsilon) f(x, y, t, \varepsilon) = 0. \tag{4.2}$$

To verify this fact it is necessary to calculate partial derivatives of the function $h(x, t, \varepsilon)$, which describes the slow integral manifold $y = h(x, t, \varepsilon)$, from the identity $G(x, h, t, \varepsilon) = 0$:

$$G_t + G_y h_t = 0, \quad G_x + G_y h_x = 0,$$

i.e. $h_t = -G_y^{-1} G_t$ and $h_x = -G_y^{-1} G_x$. Substituting these expressions into the invariance equation $\varepsilon h_t + \varepsilon h_x f = g$ we obtain

$$-\varepsilon G_y^{-1} G_t - \varepsilon G_y^{-1} G_x f = g$$

which is (4.2) under the condition $\det G_y \neq 0$.

The zero approximation to the flow on the slow integral manifold is governed by the differential-algebraic system:

$$\dot{x} = f(x, y, t, 0), \tag{4.3}$$

$$0 = g(x, y, t, 0). \tag{4.4}$$

To obtain the first order approximation, it is necessary to differentiate $g(x, y, t, \varepsilon)$, and by virtue of (4.1), the result is

$$\varepsilon \frac{d}{dt} g = g_y g + \varepsilon g_t + \varepsilon g_x f.$$

Then we equate the result to zero. As a first order approximation, the flow on the slow integral manifold is governed by the differential-algebraic system

$$\dot{x} = f(x, y, t, \varepsilon), \tag{4.5}$$

$$g_y g + \varepsilon g_t + \varepsilon g_x f = 0, \tag{4.6}$$

where terms of order $o(\varepsilon)$ can be neglected. Equation (4.6) may be written in more convenient form when $\det g_y \neq 0$

$$g + \varepsilon g_y^{-1} g_t + \varepsilon g_y^{-1} g_x f = 0. \tag{4.7}$$

In the case of an autonomous system, where g is independent of t, Eq. (4.7) takes the form

$$g + \varepsilon N = 0, \quad \det g_y \neq 0.$$

where

$$N = g_y^{-1} g_x f.$$

This is the first order approximation. We recover (4.4), the zero approximation, on setting $\varepsilon = 0$. To obtain the second order approximation, it is necessary to differentiate $g(x, y, t, \varepsilon)$ twice, using (4.1), and to equate the result to zero. The corresponding relationships are very cumbersome. Because of this we consider only the case of autonomous systems. Then, on differentiating $\varepsilon(g + \varepsilon N) = 0$ with respect to t, and noting $g_t = 0$, $N_t = 0$ the second order approximation takes the form

$$\varepsilon \frac{d}{dt}(g + \varepsilon N) = g + \varepsilon N + \varepsilon g_y^{-1} N_y g + \varepsilon^2 g_y^{-1} N_x f = 0$$

or

$$g + \varepsilon \left[N + g_y^{-1} N_y g \right] + \varepsilon^2 g_y^{-1} N_x f = 0.$$

On using $(I + \varepsilon g_y^{-1} N_y)^{-1} = I - \varepsilon g_y^{-1} N_y + O(\varepsilon)^2$ we get the result

$$g + \varepsilon N + \varepsilon^2 g_y^{-1}(N_x f - N_y N) = 0. \tag{4.8}$$

The second order approximation to the slow invariant manifold is given by (4.8) to $O(\varepsilon^2)$, and the flow on it by (4.5). In (4.5), (4.8) all terms in the expansions of f and g that lead to $O(\varepsilon^3)$ terms can be neglected.

To obtain the k-th order approximation, it is necessary to differentiate $g(x, y, t, \varepsilon)$ k times with respect to t and use (2.20).

To check these formulae it is sufficient to note that in the calculation of the asymptotic expansions of $h(x, t, \varepsilon)$

$$h = \phi + O(\varepsilon), \quad h = \phi + \varepsilon h_1 + O(\varepsilon^2),$$
$$h = \phi + \varepsilon h_1 + \varepsilon^2 h_2 + O(\varepsilon^3),$$

the use of (4.3)–(4.8) gives the same result as those given immediately before (2.24). It is necessary to use the formulae (2.20), (2.22) and (2.23) to verify this. We use Eq. (4.6) of the first order approximation in the form

$$g(x, \phi + \varepsilon h_1 + \varepsilon^2 \dots, t, \varepsilon) + \varepsilon B^{-1} g_t(x, \phi, t, 0)$$
$$+ \varepsilon B^{-1} g_x(x, \phi, t, 0) f(x, \phi, t, 0) + \varepsilon^2 \dots = 0,$$

noting that $B(x,t) = g_y = \frac{\partial g}{\partial y}$. On neglecting terms of order higher than $O(\varepsilon)$, and
noting that $g(x, \phi + \varepsilon h_1, t, \varepsilon) = \varepsilon B h_1 + \varepsilon g_1(x, \phi, t, 0) + O(\varepsilon^2)$, we get

$$B h_1 + g_1(x, \phi, t, 0) + B^{-1} g_t(x, \phi, t, 0) + B^{-1} g_x(x, \phi, t, 0) f(x, \phi, t, 0) = 0,$$

i.e.

$$h_1 = B^{-1} \left(-g_1(x, \phi, t, 0) - B^{-1} g_t(x, \phi, t, 0) - B^{-1} g_x(x, \phi, t, 0) f(x, \phi, t, 0) \right).$$

$$(4.9)$$

Taking into account that the equality $g(x, \phi, t, 0) = 0$ implies

$$g_y(x, \phi, t, 0)\phi_t + g_t(x, \phi, t, 0) = 0$$

and

$$g_y(x, \phi, t, 0)\phi_x + g_x(x, \phi, t, 0) = 0$$

we obtain $\phi_t = -B^{-1} g_t(x, \phi, t, 0)$, $\phi_x = -B^{-1} g_x(x, \phi, t, 0)$. Thus, the formulae (2.22) and (4.9) are equivalent by virtue of the fact that $B = g_y(x, \phi, t, 0)$,
$f_0 = f(x, \phi, t, 0)$. In the same way we can verify the second order approximation
and obtain from (4.8)

$$h_2 = B^{-1} \left[\frac{\partial \phi}{\partial x} f_1 + \frac{\partial h_1}{\partial x} f_0 - g_2 \right],$$

which is equivalent to (2.23) in the autonomous case.

As to the k-th order approximation, it is necessary to use the principle of
mathematical induction. We invite readers to do it for themselves.

To illustrate this approach, consider

Example 9.

$$\dot{x} = y, \qquad \varepsilon \dot{y} = x^2 + y^2 - a, \qquad a > 0. \tag{4.10}$$

Here $f = y$ and $g = x^2 + y^2 - a$. The zero approximation to the slow invariant
manifold is

$$x^2 + y^2 - a = 0.$$

The first approximation to the slow invariant manifold is

$$g + \varepsilon g_y^{-1} g_x f = y^2 + x^2 - a + \varepsilon x = 0.$$

It is easy to check that the second order approximation is

$$y^2 + (x + \varepsilon/2)^2 - a + \varepsilon^2/4 = 0$$

and it is also the exact equation for this manifold, since the function $y^2 + (x + \varepsilon/2)^2 - a + \varepsilon^2/4$ is invariant (up to an nonessential nonzero multiplier) with respect to differentiation by virtue of the differential system under consideration, i.e.

$$\varepsilon \frac{d}{dt}(y^2 + (x + \varepsilon/2)^2 - a + \varepsilon^2/4) = 2y(x^2 + y^2 - a) + \varepsilon 2(x + \varepsilon/2)y$$

$$= 2y\left(y^2 + (x + \varepsilon/2)^2 - a + \varepsilon^2/4\right).$$

This means that the second order approximation satisfies the invariance equation (4.2). The stability of the invariant manifold requires $g_y = 2y < 0$. Thus the manifold for $y > 0$ is repulsive, and that for $y < 0$ is attractive, as indicated by the arrows in Fig. 4.1.

In terms of the formalism in Chap. 2, $f = y$, $g = x^2 + y^2 - a$ and $B = g_y = 2y$. Then $x^2 + \phi^2 - a = 0$ is the zero order approximation and the second order approximation $h^2 + (x + \varepsilon/2)^2 - a + \varepsilon^2/4 = 0$ is the exact equation for $h(x, \varepsilon)$.

Example 10. The slow curve of the system

$$\frac{dx}{dt} = y,$$

$$\varepsilon \frac{dy}{dt} = by^2 + ax^2 + \alpha, \qquad (4.11)$$

in the case $ab < 0$ and $\alpha - \varepsilon^2/4 < 0$, is the hyperbola

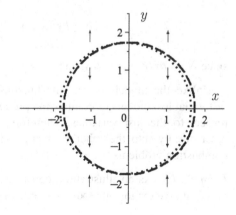

Fig. 4.1 Example 9. The slow curve (*dashed line*) and the exact slow invariant manifold (the *dotted line*) of (4.10) ($a = 3$, $\varepsilon = 0.2$) where the *arrows* indicate increasing time

Fig. 4.2 Example 10. The
slow curve (*dashed line*) and
the exact slow invariant
manifold (the *dotted line*)
of (4.11) ($a = -1$ $b = 1$,
$\alpha = -1$, $\varepsilon = 0.2$) where the
arrows indicate increasing
time

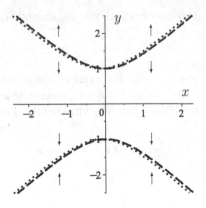

$$by^2 + ax^2 + \alpha = 0$$

the lower branch of which is attractive and the upper one is repulsive, since $g_y < 0$
($g_y > 0$) for $y < 0$ ($y > 0$) (see Fig. 4.2). The exact slow invariant manifold
is described by the equation of the second order approximation i.e., $g + \varepsilon N +$
$\varepsilon^2 g_y^{-1}(N_x f - N_y N) = 0$

$$by^2 + a\left(x + \frac{\varepsilon}{2b}\right)^2 + \alpha + \frac{\varepsilon^2 a}{4b^2} = 0,$$

since

$$\varepsilon \frac{d}{dt}\left(by^2 + a\left(x + \frac{\varepsilon}{2b}\right)^2 + \alpha + \frac{\varepsilon^2 a}{4b^2}\right)$$

$$= 2by\left(by^2 + ax^2 + \alpha + \varepsilon ax/b + \varepsilon^2 a/2b^2\right)y$$

$$= 2by\left(by^2 + a\left(x + \frac{\varepsilon}{2b}\right)^2 + \alpha + \frac{\varepsilon^2 a}{4b^2}\right),$$

since $N = ax/b, g_y = 2by$, and $f = y$.

Unlike the models of enzyme kinetics, which are linear with respect to the
fast variables, models of combustion processes usually are strongly nonlinear with
respect to the fast variables. Therefore, the use of the implicit representation
fits naturally into the scheme of approximation of slow invariant manifolds in
combustion problems.

Example 11. As an illustration, consider now the following system which is the
classical heat explosion model with reactant consumption [59, 117].

Fig. 4.3 Example 11. The slow curve (*dashed line*), the first-order (the *dotted line*) and the second-order (the *solid line*) approximations to the slow invariant manifold of (4.12) ($\varepsilon = 0.01$). Note that the approximations are not good near the critical point $\theta = 1$, where $g_\theta = 0$

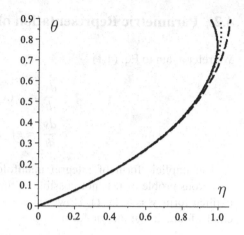

Here θ is the dimensionless temperature and η is the dimensionless concentration, and τ is the dimensionless time.

The zero-order approximation to the slow invariant manifold is

$$\frac{d\eta}{d\tau} = -\eta e^\theta,$$

$$\varepsilon \frac{d\theta}{d\tau} = \eta e^\theta - \alpha\theta. \tag{4.12}$$

Here θ is the dimensionless temperature and η is the dimensionless concentration, and τ is the dimensionless time.

The zero-order approximation to the slow invariant manifold is

$$g = \eta e^\theta - \alpha\theta = 0, \tag{4.13}$$

i.e., $\eta(\theta) = \alpha\theta e^{-\theta}$ which is stable when $\eta e^\theta - \alpha\theta < 0$. Because it is not possible to explicitly solve Eq. (4.13) for the fast variable θ, we will use the implicit form to obtain an approximation of the slow invariant manifold. The first approximation is $g + \varepsilon g_\eta f / g_\theta = 0$ i.e.,

$$\eta e^\theta - \alpha\theta - \varepsilon \eta e^{2\theta} / g_\theta = 0,$$

where $f = -\eta e^\theta, g_\theta = \eta \exp\theta - \alpha$. The second order approximation $g + \varepsilon N + \varepsilon^2 g_\theta^{-1}(N_\eta f - N_\theta N) = 0$ is

$$\eta e^\theta - \alpha\theta - \varepsilon \eta e^{2\theta} / g_\theta - \varepsilon^2 \left(\alpha\eta e^{3\theta} / g_\theta^3 + \eta^2 e^{4\theta}(\eta e^\theta - 2\alpha) / g_\theta^4\right) = 0,$$

since $N = g_\eta f / g_\theta = -\eta e^{2\theta} / g_\theta, N_\eta = \alpha e^{2\theta} / g_\theta^2, N_\theta = -\left(\eta^2 e^\theta - 2\alpha\eta\right) e^{2\theta} / g_\theta^2$, see Fig. 4.3.

We will return to this physical example in the next section and in Chap. 7.

4.2 Parametric Representation of Integral Manifolds

We refer again to Eq. (4.1)

$$\frac{dx}{dt} = f(x, y, t, \varepsilon),$$

$$\varepsilon \frac{dy}{dt} = g(x, y, t, \varepsilon). \tag{4.14}$$

The implicit form of integral manifolds has evident disadvantages, but for numerous problems it is not possible to find a solution of $g(x, y, t, 0) = 0$ in the explicit form $y = \phi(x, t)$. However, sometimes the solution of $g(x, y, t, 0) = 0$ can be found in parametric form

$$x = \chi_0(v, t), \qquad y = \varphi_0(v, t),$$

where the parameter $v \in \mathbb{R}^m$, and the following identity holds

$$g(\chi_0(v, t), \varphi_0(v, t), t, 0) \equiv 0, t \in \mathbb{R}, \quad v \in \mathbb{R}^m.$$

In this case the slow integral manifold may also be found in parametric form

$$x = \chi(v, t, \varepsilon), \qquad y = \varphi(v, t, \varepsilon),$$

where $t \in \mathbb{R}, v \in \mathbb{R}^m, \chi(v, t, 0) = \chi_0(v, t), \varphi(v, t, 0) = \varphi_0(v, t)$. The flow on the manifold is governed by an equation of the form

$$\dot{v} = F(v, t, \varepsilon), \tag{4.15}$$

and the function $F(v, t, \varepsilon)$ will be determined below. The functions χ, φ, F can be found as asymptotic expansions of the form

$$\chi(v, t, \varepsilon) = \chi_0(v, t) + \varepsilon \chi_1(v, t) + \ldots + \varepsilon^k \chi_k(v, t) + \ldots,$$

$$\varphi(v, t, \varepsilon) = \varphi_0(v, t) + \varepsilon \varphi_1(v, t) + \ldots + \varepsilon^k \varphi_k(v, t) + \ldots, \tag{4.16}$$

$$F(v, t, \varepsilon) = F_0(v, t) + \varepsilon F_1(v, t) + \ldots + \varepsilon^k F_k(v, t) + \ldots.$$

On using (4.14) and (4.15), we obtain the invariance equations

$$\frac{dx}{dt} = \frac{d\chi(v, t, \varepsilon)}{dt} = \frac{\partial \chi}{\partial t} + \frac{\partial \chi}{\partial v} F = f(\chi, \varphi, t, \varepsilon), \tag{4.17}$$

$$\varepsilon \frac{dy}{dt} = \varepsilon \frac{d\varphi(v, t, \varepsilon)}{dt} = \varepsilon \frac{\partial \varphi}{\partial t} + \varepsilon \frac{\partial \varphi}{\partial v} F = g(\chi, \varphi, t, \varepsilon) \tag{4.18}$$

and their approximations. Equating coefficients of powers of the small parameter ε we obtain equations for the zero order approximation

$$\frac{\partial \chi_0}{\partial t} + \frac{\partial \chi_0}{\partial v} F_0 = f(\chi_0, \varphi_0, t, 0), \qquad g(\chi_0, \varphi_0, t, 0) = 0,$$

for the first order approximation

$$\frac{\partial \chi_1}{\partial t} + \frac{\partial \chi_1}{\partial v} F_0 + \frac{\partial \chi_0}{\partial v} F_1 = f_x(\chi_0, \varphi_0, t, 0)\chi_1 + f_y(\chi_0, \varphi_0, t, 0)\varphi_1 + f_1,$$

$$\frac{\partial \varphi_0}{\partial t} + \frac{\partial \varphi_0}{\partial v} F_0 = g_x(\chi_0, \varphi_0, t, 0)\chi_1 + g_y(\chi_0, \varphi_0, t, 0)\varphi_1 + g_1,$$

where $f_1 = f_\varepsilon(\chi_0, \varphi_0, t, 0), g_1 = g_\varepsilon(\chi_0, \varphi_0, t, 0)$, and so on.

The two vector equations of the zero order approximation contain three unknown vector functions χ_0, φ_0 and F_0, or, what is the same, $m + n$ scalar equations contain $m + n + m$ unknown scalar functions. The same is true of the first order approximation. In general, the relationships (4.17), (4.18) contain unknown functions χ, φ, F. In a specific problem it is possible on many occasions to consider one of these functions, or any m scalar components of χ, φ and F, as known functions, and all others may be found from (4.17), (4.18). Moreover, it is possible at any step of the calculation of the coefficients in (4.16) to choose any m components of these coefficients as given functions. In the case that F is a given or known function, Eqs. (4.17), (4.18) are used to calculate the coefficients in the asymptotic expansions of χ and φ. If it is possible to predetermine the function χ, then these equations allow the calculation of F and φ. To clarify this we consider several examples.

Note that in the case of the explicit form $y = h(x, t, \varepsilon)$, we take

$$v = x, \quad \chi = v, \quad \varphi = h(v, t, \varepsilon), \quad F = f(v, h(v, t, \varepsilon), t, \varepsilon),$$

since $\dot{v} = \dot{x} = f$. Then (4.18) takes the form

$$\varepsilon \frac{\partial h}{\partial t} + \varepsilon \frac{\partial h}{\partial v} f(v, h, t, \varepsilon) = g(v, h, t, \varepsilon), \qquad h = h(v, t, \varepsilon).$$

If $\dim x = \dim y$ and the role of v is that of y, then $\varphi = v$ and

$$\frac{\partial \chi}{\partial t} + \frac{\partial \chi}{\partial v} F = f(\chi, v, t, \varepsilon), \text{and} \quad g(\chi, v, t, \varepsilon) = \varepsilon F. \tag{4.19}$$

by (4.15) and the second of (4.14) (see also (4.17)). The equation for χ follows immediately

$$\varepsilon \frac{\partial \chi}{\partial t} + \frac{\partial \chi}{\partial v} g(\chi, v, t, \varepsilon) = \varepsilon f(\chi, v, t, \varepsilon).$$

Then, under the assumption $\det(\frac{\partial \chi_0}{\partial v}) \neq 0$, it is possible to calculate χ in an asymptotic expansion. Note that $g(\chi_0, \varphi_0, t, 0) = 0$ implies that Eq. (4.15) is regularly perturbed, since the last equation in (4.19) implies, in this case, $F = O(1)$ as $\varepsilon \to 0$.

Consider Example 9 given by Eq. (4.10) for $a = 1$, viz., $\dot{x} = y$, $\quad \varepsilon \dot{y} = x^2 + y^2 - 1$. In this case the slow curve $x^2 + y^2 = 1$ may be represented in a parametric form

$$x = \cos v, \quad y = \sin v.$$

Then $\dot{x} = y \implies \dot{v} = -1$ i.e., $F(v, \varepsilon) = -1$. It is easy to find the slow invariant manifold as an asymptotic expansion. In the expansion (4.16), $\chi_0 = \cos v$ and $\varphi_0 = \sin v$ then the equations of the first order approximation are given by $-\frac{\partial \chi_1}{\partial v} = \varphi_1$ and $-\frac{1}{2} \cos v = \chi_1 \cos v - \frac{\partial \chi_1}{\partial v} \sin v$. With the particular integral $\chi_1 = -1/2$ we obtain $\varphi_1 = 0$. The second order approximation gives $-\frac{\partial \chi_2}{\partial v} = \varphi_2$ and $0 = 2\chi_2 \cos v + 2\varphi_2 \sin v + 1/4$. These two equalities imply $0 = 2\chi_2 \cos v - \sin v \frac{\partial \chi_2}{\partial v} + 1/4$, and with the particular integral $\chi_2 = -(1/8) \cos v$ we obtain $\varphi_2 = -(1/8) \sin v$. The end result then is

$$x = \cos v - \varepsilon/2 - \varepsilon^2 \frac{1}{8} \cos v + \dots,$$

$$y = \sin v - \varepsilon^2 \frac{1}{8} \sin v + \dots.$$

Note that the exact solution is

$$x = \sqrt{1 - \varepsilon^2/4} \cos v - \varepsilon/2, \quad y = \sqrt{1 - \varepsilon^2/4} \sin v.$$

Combustion models in many cases are linear with respect to the slow variables, and this permits us to use the parametric representation to find slow invariant manifolds when the fast variables play the role of v. Returning to Example 11, the combustion problem, consider the degenerate equation $\eta e^\theta - \alpha \theta = 0$ and $\eta = \chi_0(\theta) = \alpha \theta e^{-\theta}$. The role of variable v here is that of the fast variable θ. Setting $\eta = \chi_0(\theta) + \varepsilon \chi_1(\theta) + \varepsilon^2 \chi_2(\theta) + \dots$, we get

$$\varepsilon \frac{d\theta}{d\tau} = \varepsilon(\chi_1(\theta) + \varepsilon \chi_2(\theta) + \dots)e^\theta = \varepsilon F \tag{4.20}$$

and, due to (4.19),

$$\frac{\partial \chi}{\partial \theta} F = -\chi(\theta, \varepsilon)e^\theta.$$

Then, on substituting for F from (4.20), we get

$$\left(\frac{\partial\chi_0}{\partial\theta} + \varepsilon\frac{\partial\chi_1}{\partial\theta} + \varepsilon^2\frac{\partial\chi_2}{\partial\theta} + \ldots\right)(\chi_1(\theta) + \varepsilon\chi_2(\theta) + \ldots)e^\theta$$
$$= -\left(\chi_0(\theta) + \varepsilon\chi_1(\theta) + \varepsilon^2\chi_2(\theta) + \ldots\right)e^\theta.$$

On equating powers of ε, we find $\frac{\partial\chi_0}{\partial\theta}\chi_1 = -\chi_0$, $\frac{\partial\chi_0}{\partial\theta}\chi_2 + \frac{\partial\chi_1}{\partial\theta}\chi_1 = -\chi_1$ and $\frac{\partial\chi_0}{\partial\theta} = \alpha(1-\theta)e^{-\theta}$, and it is then easy to calculate $\chi_1 = \frac{\theta}{\theta-1}$ and $\chi_2 = e^\theta\frac{\theta^2(\theta-2)}{\alpha(\theta-1)^4}$.
Thus, we obtain

$$\eta = \chi(\theta,\varepsilon) = \alpha\theta e^{-\theta} + \varepsilon\frac{\theta}{\theta-1} + \varepsilon^2 e^\theta\frac{\theta^2(\theta-2)}{\alpha(\theta-1)^4} + O(\varepsilon^3).$$

This representation is correct outside some neighborhood of $\theta = 1$, and it gives the approximation of the attractive (repulsive) one-dimensional slow invariant manifold if $0 \le \theta < 1$ $(1 < \theta)$.

Finally we consider the third order differential equation which has the form of a differential system with two slow and one fast variable

$$\dot{x}_1 = x_2, \quad \dot{x}_2 = y, \quad \varepsilon\dot{y} = -y - e^y - x_1 - x_2.$$

Note that this system possesses an attractive slow invariant manifold since $\frac{\partial}{\partial y}(-y - e^y - x_1 - x_2) = -1 - e^y < 0$. The degenerate equation $0 = -y - e^y - x_1 - x_2$ cannot be solved with respect to the fast variable y, but it can be solved with respect to one of the slow variables x_1 or x_2. Thus, the fast variable y and the slow variable x_2 may be chosen as parameters and the slow invariant manifold will be represented in the form $x_1 = \chi(x_2, y, \varepsilon) = \chi_0(x_2, y) + \varepsilon\chi_1(x_2, y) + \varepsilon^2\chi_1(x_2, y) + O(\varepsilon^3)$, where $\chi_0(x_2, y) = -y - e^y - x_2$. The flow on this manifold is described by the equations

$$\dot{x}_2 = y, \quad \varepsilon\dot{y} = -\chi_1(x_2, y) - \varepsilon\chi_1(x_2, y) + O(\varepsilon^2)$$

and the invariance equation

$$\frac{\partial\chi}{\partial x_2}y + \frac{\partial\chi}{\partial y}\frac{1}{\varepsilon}(-y - e^y - \chi - x_2) = x_2$$

takes the form

$$\left(\frac{\partial\chi_0}{\partial x_2} + \varepsilon\frac{\partial\chi_1}{\partial x_2} + \ldots\right)y + \left(\frac{\partial\chi_0}{\partial y} + \varepsilon\frac{\partial\chi_1}{\partial y} + \ldots\right)(-\chi_1 - \varepsilon\chi_2 - \ldots) = x_2.$$

Equating the powers of ε we obtain

$$\chi_1 = \frac{x_2 + y}{1 + e^y}, \quad \chi_2 = \left(-\frac{\partial \chi_1}{\partial x_2} y + \frac{\partial \chi_1}{\partial y} \chi_1\right) / (1 + e^y)$$

after taking into account that $\frac{\partial \chi_0}{\partial y} = -1 - e^y$.

Chapter 5
Singular Singularly Perturbed Systems

Abstract In this chapter we consider singularly perturbed differential systems whose degenerate equations have an isolated but not simple solution. In that case, the standard theory to establish a slow integral manifold near this solution does not work. Applying scaling transformations and using the technique of gauge functions we reduce the original singularly perturbed problem to a regularized one such that the existence of slow integral manifolds can be established by means of the standard theory. We illustrate the method by several examples from control theory and chemical kinetics.

5.1 Introduction

For a better idea of the problems we wish to examine, and to gain some insight into why the term in the title is used, we initially consider the following differential system

$$\varepsilon \dot{z}_1 = 2z_1 - z_2, \quad \varepsilon \dot{z}_2 = (6 + 3\varepsilon)z_1 - 3z_2,$$

or, in the vector form

$$\varepsilon \begin{pmatrix} \dot{z}_1 \\ \dot{z}_2 \end{pmatrix} = A(\varepsilon) \begin{pmatrix} z_1 \\ z_2 \end{pmatrix}, \text{ where } A(\varepsilon) = \begin{pmatrix} 2 & -1 \\ 6 + 3\varepsilon & -3 \end{pmatrix}.$$

At first glance it would seem that there are two fast variables z_1 and z_2 and we apply the proposed approach to the analysis of this system. Setting ε equal to zero we obtain the linear algebraic system

$$2z_1 - z_2 = 0, \quad 6z_1 - 3z_2 = 0.$$

Apart from the trivial solution this system possesses an one-parameter family of solutions $z_1 = s$, $z_2 = 2s$, where s is a real parameter. Thus there is no isolated solution to the degenerate system. The reason is that the matrix is *singular*, i.e. $\det A(0) = 0$ and the singularly perturbed system in this case is called a *singular singularly perturbed system*. In fact, in this particular system it is possible to extract

© Springer International Publishing Switzerland 2014

93

E. Shchepakina et al., *Singular Perturbations*, Lecture Notes in Mathematics 2114, DOI 10.1007/978-3-319-09570-7_5

a slow variable and obtain a system with one slow and one fast variable. Taking into account that the rows of matrix A are proportional for $\varepsilon = 0$ (proportionality constant is equal 3), we introduce a new variable $x = z_2 - 3z_1$, and obtain the following differential equation for the slow variable $\dot{x} = -x + z_2$. To obtain the full solution, it is possible to use either of the two equations for z_1 or z_2 as a fast equation. If we choose the equation for z_2, then, after taking into account $x - z_2 = -3z_1$, we obtain the singularly perturbed equation $\varepsilon \dot{z}_2 = -(2 + \varepsilon)x - (1 - \varepsilon)z_2$. As a result we obtain the system

$$\dot{x} = -x + z_2, \quad \varepsilon \dot{z}_2 = -(2 + \varepsilon)x - (1 - \varepsilon)z_2$$

which has the form (2.2). It is easy to check by direct substitution into the invariance equation that this last system possesses the one-dimensional attractive slow invariant manifold $z_2 = kx$. On substituting for z_2 the above equations imply

$$\varepsilon k(-1 + k)x = -(2 + \varepsilon)x - (1 - \varepsilon)kx,$$

and this implies

$$\varepsilon k^2 + (1 - 2\varepsilon)k + 2 + \varepsilon = 0.$$

Setting in the last equation $k = k_0 + \varepsilon k_1 + O(\varepsilon^2)$ and equating the powers of ε, we obtain $k_0 = -2$, $k_1 = -9$. Thus, the invariant manifold has the form

$$z_2 = -(2 + 9\varepsilon + O(\varepsilon^2))x = -(2 + 9\varepsilon + O(\varepsilon^2))(z_2 - 3z_1),$$

or, in equivalent form

$$z_2 = (2 + 3\varepsilon + O(\varepsilon^2))z_1. \tag{5.1}$$

A singularly perturbed differential system can often be written in the form

$$\varepsilon \dot{z} = Z(z, t, \varepsilon), \qquad z \in \mathbb{R}^{m+n}, \quad t \in \mathbb{R}, \tag{5.2}$$

$$z = (z_1, z_2, \ldots, z_{m+n}),$$

$$Z = (Z_1, Z_2, \ldots, Z_{m+n}); \ Z_i = Z_i(z, t, \varepsilon),$$

where $0 \le \varepsilon \ll 1$, and the vector-function Z is sufficiently smooth. Suppose that for $\varepsilon = 0$ the limit system $Z(z, t, 0) = 0$ has a family of solutions

$$z = \psi(v, t), \qquad v \in \mathbb{R}^m, \quad t \in \mathbb{R}, \tag{5.3}$$

with a sufficiently smooth vector-function ψ . If $m > 0$ then (5.2) is called as *singular singularly perturbed system* [66]. In the example just given $m + n = 2$ and $m = 1$.

We try to find a slow integral manifold of (5.2)

$$z = P(v, t, \varepsilon), \tag{5.4}$$

with a flow described by an equation of the form

$$\dot{v} = Q(v, t, \varepsilon). \tag{5.5}$$

We shall restrict our consideration to smooth integral surfaces situated in the ε-neighborhood of the slow surface $z = \psi(v, t)$, i. e.

$$P(v, t, 0) = \psi(v, t),$$

and to the motion on the integral surface which is described by differential equations of form (5.5) with a smooth right hand side. Consider, for example, the linear homogeneous singularly perturbed differential system

$$\varepsilon \frac{dz}{dt} = Az \tag{5.6}$$

with a constant $(m+n) \times (m+n)$ matrix A. If A is a nonsingular matrix then the degenerate equation $Ax = 0$ has a unique solution $x = 0$, but in the case of singular matrix A the degenerate equation $Ax = 0$ has an m-parameter family of solutions $(m = \dim x - \text{rank} A)$ [98, 192]. In this case it is reasonable to call (5.6) a *singular singularly perturbed* differential system [66, 83].

5.2   Construction of Slow Integral Manifold

Suppose the following hypotheses hold for the system (5.2):

(*i*) the rank of the matrix $\psi_v(v, t)$ is equal to m;
(*ii*) the rank of the matrix $A(v, t) = Z_z(\psi(v, t), t, 0)$ is equal to n where $Z_z = \left(\frac{\partial Z_i}{\partial z_j} \right) (i, j = 1, \ldots, n + m)$;
(*iii*) the matrix $A(v, t)$ has an m-fold zero eigenvalue and n other eigenvalues $\lambda_i(v, t)$ that satisfy the inequality

$$\text{Re}\lambda_i(v, t) \leq -2\alpha < 0, \qquad t \in \mathbb{R}, \qquad v \in \mathbb{R}^m. \tag{5.7}$$

Differentiation of $Z(\psi(v, t), t, 0) = 0$ with respect to v gives

$$A(v, t)\psi_v(v, t) = 0.$$

This means, by (*iii*) above, that the $(m + n) \times (m + n)$-matrix $A(v, t)$ possesses m linearly independent eigenvectors, which are columns of $\psi_v(v, t)$, corresponding to multiple zero eigenvalues [98, 192].

We denote by $(*)^T$ the transpose of a matrix $(*)$. Let D_1^T be an $(m+n) \times n$-matrix, the columns of which are n linearly independent eigenvectors corresponding to m zero eigenvalues and D_2^T be such an $(m+n) \times m$-matrix so that (D_1^T, D_2^T) is a non-singular $(m + n) \times (m + n)$-matrix, i.e. $\det(D_1^T, D_2^T) \neq 0$. Then $A^T(D_1^T, D_2^T) = (0 \ B^T)$, due to the zero eigenvalues of D_1^T, where B is non-singular. In other words

$$DA = \begin{pmatrix} 0 \\ B \end{pmatrix}, \text{ for } D = \begin{pmatrix} D_1 \\ D_2 \end{pmatrix}.$$

Thus, the result of multiplying the non-singular matrix D on the right by A provides the zero $m \times (m + n)$-block and the non-singular $n \times (m + n)$-block B.

The rank of B is equal to n. Consequently, without loss of generality, the system (5.2) with renumbered variables and equations may be considered to be of the form (the model of bimolecular reaction as a 3D example is considered below)

$$\varepsilon\dot{x} = f_1(x, y_2, t, \varepsilon), \qquad x \in \mathbb{R}^m, \tag{5.8}$$

$$\varepsilon\dot{y}_2 = f_2(x, y_2, t, \varepsilon), \qquad y_2 \in \mathbb{R}^n, \tag{5.9}$$

when the following assumptions hold.

(**B₁**).   The equation $f_2(x, y_2, t, 0) = 0$ has a smooth isolated root $y_2 = \varphi(x, t)$ with $x \in \mathbb{R}^m, t \in \mathbb{R}$, and $f_2(x, \varphi(x, t), t, 0) = 0$.
(**B₂**).   The Jacobian matrix

$$A(x, t) = \begin{pmatrix} f_{1x} & f_{1y_2} \\ f_{2x} & f_{2y_2} \end{pmatrix}\Bigg|_{y_2 = \varphi(x, t), \, \varepsilon = 0}$$

on the surface $y_2 = \varphi(x, t)$ has an m-fold zero eigenvalue and m linearly independent eigenvectors, and the matrix $B(x, t) = f_{2y_2}(x, \varphi(x, t), t, 0)$ has n eigenvalues satisfying (5.7) where $v = x$.
(**B₃**).   In the domain

$$\Omega = \{(x, y_2, t, \varepsilon) \mid x \in \mathbb{R}^m, \ ||y_2 - \varphi(x, t)|| \leq \rho, \ t \in \mathbb{R}, \ 0 \leq \varepsilon \leq \varepsilon_0\},$$

the functions f_1, f_2 and the matrix A are continuously differentiable $(k + 2)$ times $(k \geq 0)$ for some positive ε_0 and ρ.

Using the change of variable $y_2 = y_1 + \varphi(x, t)$ in (5.8), (5.9), we obtain the following equations for x and y_1

$$\varepsilon\dot{x} = C(x, t)y_1 + F_1(x, y_1, t) + \varepsilon X(x, y_1, t, \varepsilon), \tag{5.10}$$

$$\varepsilon\dot{y}_1 = B(x, t)y_1 + F_2(x, y_1, t) + \varepsilon Y(x, y_1, t, \varepsilon), \tag{5.11}$$

where

$$C(x,t) = f_{1y_2}(x,\varphi(x,t),t,0), \qquad B(x,t) = f_{2y_2}(x,\varphi(x,t),t,0),$$
$$F_1(x,y_1,y) = f_1(x,y_1+\varphi(x,t),t,0) - C(x,t)y_1,$$
$$F_2(x,y_1,t) = f_2(x,y_1+\varphi(x,t),t,0) - B(x,t)y_1,$$
$$\varepsilon X(x,y_1,t,\varepsilon) = f_1(x,y_1+\varphi(x,t),t,\varepsilon) - f_1(x,y_1+\varphi(x,t),t,0),$$
$$\varepsilon Y(x,y_1,t.\varepsilon) = f_2(x,y_1+\varphi(x,t),t,\varepsilon) - f_2(x,y_1+\varphi(x,t),t,0).$$

Note that the vector-functions F_i $(i = 1,2)$ satisfy the relations $\|F_i(x,y_1,t)\| = O(\|y_1\|^2)$. Thus, $\varepsilon^{-1}F_i(x,\varepsilon y,t)$ are continuous in Ω.

If the assumptions $(\mathbf{B_1})$–$(\mathbf{B_3})$ are satisfied then there exists ε_1, $0 < \varepsilon_1 < \varepsilon_0$, such that for any $\varepsilon \in (0,\varepsilon_1)$ the system (5.10), (5.11) possesses a unique slow integral manifold $y_1 = \varepsilon p(x,t,\varepsilon)$. On this manifold the flow of the system is governed by the equation

$$\dot{x} = X_1(x,t,\varepsilon),$$

where $X_1(x,t,\varepsilon) = C(x,t)p(x,t,\varepsilon) + X(x,\varepsilon p,t,\varepsilon) + \varepsilon^{-1}F_1(x,\varepsilon p,t)$, and the function $p(x,t,\varepsilon)$ is k times continuously differentiable with respect to x and t.

Note that the change of variable $y_1 = \varepsilon y$ converts the system (5.10), (5.11) to the form

$$\dot{x} = \tilde{X}(x,y,t,\varepsilon), \qquad x \in \mathbb{R}^m, \quad t \in \mathbb{R}, \tag{5.12}$$
$$\varepsilon \dot{y} = \tilde{Y}(x,y,t,\varepsilon), \qquad y \in \mathbb{R}^n, \tag{5.13}$$

where $\tilde{X}(x,y,t,\varepsilon) = C(x,t)y + \varepsilon^{-1}F_1(x,\varepsilon y,t) + X(x,\varepsilon y,t,\varepsilon)$, and $\tilde{Y}(x,y,t,\varepsilon) = B(x,t)y + \varepsilon^{-1}F_2(x,\varepsilon y,t) + Y(x,\varepsilon y,t,\varepsilon)$, since the vector-functions F_i $(i = 1,2)$ satisfy the relations $\|F_i(x,y_1,t)\| = O(\|y_1\|^2)$, and the role of the slow variable is now played by x.

We return now to the model of bimolecular reaction (3.31)–(3.33) and multiply both sides of the first equation by ε, to obtain the differential system in the form (5.2). Setting $x_1 = z_1$, $y = z_3$, $z = z_2$ we get

$$\varepsilon \frac{dz_1}{dt} = \varepsilon(1 - k_2 z_1 z_3) = Z_1(z_1,z_2,z_3), \tag{5.14}$$

$$\varepsilon \frac{dz_2}{dt} = -2\varepsilon k_1 z_2^2 + 2z_3 + \varepsilon(k_2 z_1 z_3 - z_2) = Z_2(z_1,z_2,z_3,\varepsilon), \tag{5.15}$$

$$\varepsilon \frac{dz_3}{dt} = \varepsilon k_1 z_2^2 - z_3 = Z_3(z_1,z_2,z_3,\varepsilon). \tag{5.16}$$

For $\varepsilon = 0$ the limit system

$$0 = 0, \quad 0 = 2z_3, \quad 0 = -z_3$$

has the family of solutions $z_1 = v_1$, $z_2 = v_2$, $z_3 = 0$. In this case $m + n = 3$ and $m = 2$, and the matrix A for the limit system takes the form

$$A = \begin{pmatrix} 0 & 0 & 0 \\ 0 & 0 & 2 \\ 0 & 0 & -1 \end{pmatrix}.$$

This matrix and the matrix A^T have a double zero eigenvalue and the corresponding eigenvectors of A^T are

$$\begin{pmatrix} 1 \\ 0 \\ 0 \end{pmatrix}, \begin{pmatrix} 0 \\ 1 \\ 2 \end{pmatrix}.$$

Thus, D^T can be chosen in the form

$$D^T = \begin{pmatrix} 1 & 0 & 0 \\ 0 & 1 & 0 \\ 0 & 2 & 1 \end{pmatrix}, \quad i.e. \quad D = \begin{pmatrix} 1 & 0 & 0 \\ 0 & 1 & 2 \\ 0 & 0 & 1 \end{pmatrix}.$$

and the transformation of (5.14)–(5.16), which reduces this system to the form (3.34)–(3.36) is equivalent to the multiplication of the vector of derivatives and the vector of the right hand sides of the differential system (3.31)–(3.33) by the matrix D. As a result we obtain the system

$$\varepsilon \frac{dz_1}{dt} = Z_1(z_1, z_2, z_3),$$

$$\varepsilon \left(\frac{dz_2}{dt} + 2\frac{dz_3}{dt} \right) = Z_2(z_1, z_2, z_3, \varepsilon) + 2Z_3(z_1, z_2, z_3, \varepsilon),$$

$$\varepsilon \frac{dz_3}{dt} = Z_3(z_1, z_2, z_3, \varepsilon).$$

Introducing the new variable x_2 by $x_2 = z_2 + 2z_3$ and setting $z_1 = x_1$, $z_3 = y$ leads to the system (3.34)–(3.36). This means that we construct the same transformation of the differential system under consideration which was used in Sect. 3.3.2.

5.3   Implicit Slow Integral Manifolds

In the previous chapter we pointed out that in many cases slow integral manifolds may be obtained in an implicit form. We use this approach for the system (5.2) in the case

$$\varepsilon \dot{x} = X(x, y, \varepsilon) = Q(x, y) + \varepsilon q(x, y, \varepsilon), \tag{5.17}$$

$$\varepsilon \dot{y} = Y(x, y, \varepsilon) = KQ(x, y) + \varepsilon p(x, y, \varepsilon). \tag{5.18}$$

We require

$$Y(x, y, 0) = KX(x, y, 0)$$

and this equality means that components of $Y(x, y, 0)$ can be represented as a linear combination of components of $X(x, y, 0)$, where the matrix K is formed from the coefficients of these linear combinations. This situation is typical of a wide class of chemical kinetics systems [96].

Introducing a new variable $v = y - Kx$, we obtain the following differential equation for the slow variable

$$\dot{v} = p(x, y, \varepsilon) - Kq(x, y, \varepsilon).$$

Suppose $\det(Q_x + Q_{yK}) \neq 0$. This condition is necessary to satisfy $(\mathbf{B_1})$ and $(\mathbf{B_2})$. To obtain the full solution, it is possible to use either the equation for x or for y as a fast equation. If we use the equation for x as the slow subsystem we then obtain

$$\dot{v} = p(x, v + Kx, \varepsilon) - Kq(x, v + Kx, \varepsilon),$$

$$\varepsilon \dot{x} = Q(x, v + Kx) + \varepsilon q(x, v + Kx, \varepsilon).$$

In this case the slow invariant manifold can be obtained in an implicit form. The zeroth approximation of the slow invariant manifold is given by $Q(x, v + Kx) = 0$. To obtain the first order approximation, it is necessary to differentiate $Q(x, v + Kx) + \varepsilon q(x, v + Kx, \varepsilon)$ with respect to t

$$\frac{d}{dt} (Q(x, v + Kx) + \varepsilon q(x, v + Kx, \varepsilon))$$

$$= \frac{\partial}{\partial v} (Q(x, v + Kx) + \varepsilon q(x, v + Kx, \varepsilon)) \dot{v}$$

$$+ \frac{\partial}{\partial x} (Q(x, v + Kx) + \varepsilon q(x, v + Kx, \varepsilon)) \dot{x},$$

and use the equations for $\dot{v}$ and $\varepsilon \dot{x}$ above. As a first approximation, the flow on the slow invariant manifold is governed by the differential-algebraic system

$$\dot{v} = p(x, v + Kx, \varepsilon) - Kq(x, v + Kx, \varepsilon), \tag{5.19}$$

$$(Q_x + \varepsilon q_x + Q_y K + \varepsilon q_y K)(Q + \varepsilon q) + \varepsilon Q_y(p - Kq) = 0, \tag{5.20}$$

where terms of order $o(\varepsilon)$ can be neglected. Here Eq. (5.20) describes approximately the slow invariant manifold.

Now we will obtain the same approximation directly from the original differential system by differentiating the function $Q + \varepsilon q$ with respect to t and using the original system (5.17), (5.18):

$$(Q_x + \varepsilon q_x)(Q + \varepsilon q) + (Q_y + \varepsilon q_y)(KQ + \varepsilon p) = 0. \qquad (5.21)$$

If we neglect terms of order $o(\varepsilon)$, then Eqs. (5.20), (5.21) take the form

$$(Q_x + Q_y K)Q + \varepsilon(q_x + q_y K)Q + \varepsilon(Q_x q + Q_y p) = 0, \qquad (5.22)$$

or

$$(Q_x + \varepsilon q_x)(Q + \varepsilon q) + (Q_y + \varepsilon q_y)(KQ + \varepsilon p) = 0. \qquad (5.23)$$

The last equation is just

$$X_x X + X_y Y = 0, \qquad (5.24)$$

where X and Y are given in (5.17) and (5.18). This implies that we do not need to know the matrix K to obtain the first order approximation to the slow invariant manifold.

As an example, we return to that considered at the beginning of the chapter. We have $X = 2z_1 - z_2$, $Y = 3(2z_1 - z_2) + 3\varepsilon z_1$, i.e. $Q = 2z_1 - z_2$, $q = 0$, $K = 3$, $p = 3z_1$. Then (5.24) takes the form $2z_1 - z_2 + \varepsilon 3z_1 = 0$ which is equivalent to (5.1) with an accuracy $O(\varepsilon^2)$.

5.4   Parametric Representation of Integral Manifolds

As mentioned above, sometimes the slow integral manifold can be found as a parametric function.

Returning to the system (5.2), $\varepsilon \dot{z} = Z(z, t, \varepsilon)$, we use the parametric form to describe the slow integral manifold and the flow on the manifold, i.e.

$$z = P(v, t, \varepsilon), \qquad \dot{v} = Q(v, t, \varepsilon),$$

see (5.4), (5.5). The functions P and Q will be found as asymptotic expansions

$$P(v, t, \varepsilon) = P_0(v, t) + \varepsilon P_1(v, t) + \ldots + \varepsilon^k P_k(v, t) + \ldots,$$

$$Q(v, t, \varepsilon) = Q_0(v, t) + \varepsilon Q_1(v, t) + \ldots + \varepsilon^k Q_k(v, t) + \ldots,$$

where $P_0(v, t) = \psi(v, t)$ by (5.3). Differentiating P with respect to t, and using (5.2), (5.5), gives

$$\varepsilon \frac{dP}{dt} = \varepsilon \frac{\partial P}{\partial t} + \varepsilon \frac{\partial P}{\partial v} Q = Z(P, t, \varepsilon). \tag{5.25}$$

Write the Taylor series expansion of $Z(P, t, \varepsilon)$ about $\varepsilon = 0$ as

$$Z(P, t, \varepsilon) = Z(P_0, t, 0) + \varepsilon \Theta_1(P_0, P_1, t)$$

$$+ \ldots + \varepsilon^k \Theta_k(P_0, P_1, \ldots, P_k, t) \ldots,$$

and represent Θ_k $(k \geq 1)$ in the form

$$\Theta_k(P_0, \ldots, P_k, t) = Z_P(P_0, t, 0) P_k + R_k(P_0, P_1, \ldots, P_{k-1}, t),$$

where $Z_P = \frac{\partial Z}{\partial P}$. In particular,

$$\Theta_1(P_0, P_1, t) = Z_P(P_0, t, 0) P_1 + R_1,$$

where $R_1 = Z_\varepsilon(P_0, t, 0)$. Equating powers of ε, we obtain from (5.25) with $\varepsilon = 0$

$$Z(P_0, t, 0) = 0.$$

In keeping with (5.3), let $P_0(v, t) = \psi(v, t)$.

Using the notation $A(v, t) = Z_z(\psi(v, t), t, 0)$ and on using (5.2) and the form for Θ_1 above, we obtain at order ε

$$\frac{\partial \psi}{\partial t} + \frac{\partial \psi}{\partial v} Q_0 = A P_1 + R_1. \tag{5.26}$$

Equation (5.26) contains two unknown functions P_1 and Q_0. Where P_1 is concerned, Eq. (5.26) may be considered as a nonhomogeneous linear algebraic system with a singular matrix, $\det A(v, t) \equiv 0$, $v \in \mathbb{R}^m$, $t \in \mathbb{R}$. Thus, Q_0 is needed to ensure the solvability of the system. It is apparent that we have some freedom in choosing the form of Q_0 and P_1. To determine these functions uniquely, we multiply equation (5.26) on the left by the matrix D, introduced in Sect. 5.2, and obtain the pair of equations

$$D_1 \frac{\partial \psi}{\partial t} + D_1 \frac{\partial \psi}{\partial v} Q_0 = D_1 R_1, \tag{5.27}$$

since $D_1 A = 0$, and

$$D_2 \frac{\partial \psi}{\partial t} + D_2 \frac{\partial \psi}{\partial v} Q_0 = B P_1 + D_2 R_1, \tag{5.28}$$

since $D_2 A = B$. If it is assumed additionally that the matrix

$$D_1 = \partial\psi/\partial v$$

is invertible, (5.27) gives

$$Q_0 = (D_1\psi_v)^{-1}D_1(R_1 - \psi_t),$$

and this permits us to determine P_1 uniquely from (5.28): $P_1 = B^{-1}D_2(\psi_t + \psi_v Q_0 - R_1)$. The determination of the pairs of later coefficients P_k, Q_{k-1} is carried out in the same way.

As a generalization of the example discussed in Sect. 5.3 consider the system of two scalar equations

$$\varepsilon\dot{z}_1 = f(z_1, z_2, t, \varepsilon) + \varepsilon f_1(z_1, y, t, \varepsilon),$$

$$\varepsilon\dot{z}_2 = kf(z_1, z_2, t, \varepsilon) + \varepsilon f_2(z_1, z_2, t, \varepsilon).$$

Setting $\varepsilon = 0$ we obtain two equivalent equations $f(z_1, z_2, t, 0) = 0$ and $kf(z_1, z_2, t, 0) = 0$.

Introducing a new variable $x = z_2 - kz_1$, we obtain the following differential equation for the slow variable

$$\dot{x} = f_2(z_1, z_2, t, \varepsilon) - kf_1(z_1, z_2, t, \varepsilon).$$

To obtain the full solution, it is possible to use either of the two equations for z_1 or z_2 as a fast equation.

5.5   High-Gain Control

We consider a nonlinear control system

$$\dot{x} = f(x) + B(x)u, \quad x(0) = x_0, \tag{5.29}$$

with, in general, a nonlinear vector function $f(x)$ and matrix function $B(x)$, and high-gain feedback

$$u = -\frac{1}{\varepsilon}KS(x), \tag{5.30}$$

where $x \in \mathbb{R}^n$, $u \in \mathbb{R}^r$, $t \geq 0$, K is a constant $r \times m$-matrix and ε is a small positive parameter, see [201, 202]. This control problem is of practical significance in itself and is of theoretical significance in the theory of variable structure control systems and the equivalent control method [92, 201, 202]. For linear control systems this problem was analyzed in [92, 128, 216]. The stabilization of movements of a

mechanical system along a surface is a known problem [106, 202]. It often happens
it is necessary that the working body moves along the given surface. It can be the
movement of an electrode of the welding manipulator along a seam, the movement
of a ladle of a dredge along a planned surface of the ground, movement of the cutting
tool of the manufacturing machine along a processed surface, etc.

The vector function f and the matrix function B are taken to be sufficiently
smooth and bounded. The control vector u is to be selected in such a way as to
transfer the vector x from $x = x_0$ to a sufficiently small neighborhood of a smooth
m-dimensional surface $S(x) = 0$.

Suppose that we can choose the matrix K in such a way that the matrix
$-N(x, t) = -GBK$ is stable[1] and its inverse matrix is bounded, where $G(x) =
\partial S / \partial x$. We introduce the additional variable $y = S(x)$. Substituting for u
from (5.30) into the original Eq. (5.29) and noting that $\dot{y} = S'(x)\dot{x}$, x and y satisfy
the system

$$\varepsilon \dot{x} = \varepsilon f(x) - B(x)Ky, \quad x(0) = x_0,$$

$$\varepsilon \dot{y} = \varepsilon G(x)f(x) - N(x)y, \quad y(0) = y_0 = S(x_0).$$

The reduced $(\varepsilon = 0)$ algebraic problem possesses an n-parameter family of
solutions $x = v$, $y = 0$. The role of the matrix $A(v, t)$ in Sect. 5.2 is played
by the singular matrix

$$\begin{pmatrix} 0 & -BK \\ 0 & -N \end{pmatrix}.$$

The singular singularly perturbed differential system above possesses an n-
dimensional slow integral manifold

$$x = v, \quad y = \varepsilon N^{-1}(v)G(v)f(v) + O(\varepsilon^2).$$

The flow on the manifold is governed by

$$\dot{v} = [I - B(v)KN^{-1}(v)G(v)]f(v) + O(\varepsilon).$$

Introduce the new variables

$$x = v + B(v)KN^{-1}(v)z; \quad y = z + \varepsilon N^{-1}(x)G(x)f(x).$$

Then we obtain the equations

$$\dot{v} = (I - BKN^{-1}G)f + O(\varepsilon), \quad \varepsilon \dot{z} = -(N + O(\varepsilon))z$$

---

[1]A stable matrix is one whose eigenvalues all have strictly negative real parts.

for v and z. By virtue of the result in (2.7) and (2.8) it is clear that the representations

$$x = v + O(e^{-vt/\varepsilon}), \quad y = \varepsilon\varphi(v, \varepsilon) + O(e^{-vt/\varepsilon}),$$

where $\varphi = N^{-1}Gf + O(\varepsilon)$, are valid for some $v > 0$ for all $t > 0$. Thus, under the given control law

$$u = -\frac{1}{\varepsilon}KS(x),$$

the trajectory $x(t)$ very quickly attains the ε-neighborhood of $S(x) = 0$.

It is easy to see, as suggested in [176], that the modified control

$$u = -\frac{1}{\varepsilon}K\left[S(x) + \varepsilon N^{-1}(x)G(x)f(x)\right],$$

with the stable matrix $-N(x) = -GBK$, is preferable, because it guides the trajectory of x in a time Δt to the $e^{-v\Delta t\varepsilon^{-1}}$-neighborhood of $S(x) = 0$. To verify this note that under this control for the variable x we obtain the equation

$$\varepsilon\dot{x} = \varepsilon\left[I - B(x)K(G(x)B(x)K)^{-1}G(x)\right]f(x) - B(x)KS(x),$$

and for the variable $y = S(x)$ using the identity

$$
\begin{aligned}
\dot{y} = \frac{dS(x)}{dt} &= \frac{\partial S}{\partial x}\dot{x} = G(x)\dot{x} \\
&= G(x)\left[I - B(x)K(G(x)B(x)K)^{-1}G(x)\right]f(x) - \varepsilon^{-1}G(x)B(x)KS(x) \\
&= \left[G(x) - (G(x)B(x)K)(G(x)B(x)K)^{-1}G(x)\right]f(x) - \varepsilon^{-1}G(x)B(x)KS(x) \\
&= -\varepsilon^{-1}G(x)B(x)KS(x) = -\varepsilon^{-1}N(x)y,
\end{aligned}
$$

we obtain the equation

$$\varepsilon\dot{y} = -N(x)y.$$

Then

$$y = O\left(e^{-v\varepsilon^{-1}t}\right), \quad v > 0, \ t > 0, \ \varepsilon \to 0.$$

To explain the last relationship we use inequalities (2.8) in the case of $h \equiv 0$. This relationship is obvious in the simple case of scalar y and constant $N > 0$ with $v = N$.

Additionally we note the following. Since the matrix $-N$ is stable (see the footnote at the beginning of this section) we can claim that the solution of the equation

Fig. 5.1 Example 12. The parabola and two trajectories corresponding to high-gain control law (*the dashed line*) and modified control law (*the solid line*)

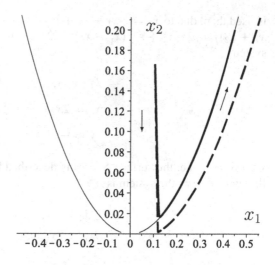

$$\varepsilon \dot{y} = -Ny, \quad y(0) = y_0$$

satisfies the inequality

$$y \leq N_0 \|y_0\| e^{-vt/\varepsilon}, t > 0$$

for some positive numbers v and N_0. All the details can be found in the Appendix; see the inequality (9.23) there.

By way of illustration we consider the following control system (Fig. 5.1):

Example 12.

$$\dot{x}_1 = x_2,$$

$$\dot{x}_2 = -x_1 - x_2 + u,$$

with $S : x_1^2 - x_2 = 0$. In this case $n = 2, m = 1, r = 1$,

$$f = \begin{pmatrix} x_2 \\ -x_1 - x_2 \end{pmatrix}, \quad B = \begin{pmatrix} 0 \\ 1 \end{pmatrix}, \quad \partial S/\partial x = G = (2x_1, \; -1),$$

and $N = GBK = -K$, where K is a scalar since GB and S are scalar. Setting $K = -1$ we obtain $N = 1$.

If the modified control law

$$u = \varepsilon^{-1} \left(x_1^2 - x_2 + \varepsilon(2x_1 x_2 + x_1 + x_2) \right)$$

is used then, due to $\dot{x}_2 = -x_1 - x_2 + u = -x_1 - x_2 + \varepsilon^{-1}(x_1^2 - x_2 + \varepsilon(2x_1x_2 + x_1 + x_2)) = 2x_1x_2 + \varepsilon^{-1}(x_1^2 - x_2) = 2x_1x_2 + \varepsilon^{-1}S(x)$, we obtain the following system

$$\dot{x}_1 = x_2,$$
$$\dot{x}_2 = 2x_1x_2 + \varepsilon^{-1}y,$$
$$\dot{y} = -\varepsilon^{-1}y$$

with $S = y$, i.e. the surface $S = 0$ is described by the equation $z = 0$. It is clear that with $y = \varepsilon z$ the system is

$$\dot{x}_1 = x_2,$$
$$\dot{x}_2 = 2x_1x_2 + z,$$
$$\varepsilon\dot{z} = -z,$$

and it possesses the attractive slow invariant manifold $z = 0$ since $\varepsilon\dot{z} = g = -z$ and $\partial g/\partial z = -1$. It should be noted that, with the modified control law $u = \varepsilon^{-1}(x_1^2 - x_2 + \varepsilon(2x_1x_2 + x_1 + x_2))$ currently selected, the control system

$$\dot{x}_1 = x_2,$$
$$\dot{x}_2 = 2x_1x_2 + \varepsilon^{-1}(x_1^2 - x_2)$$

has the attractive slow invariant manifold $x_2 = x_1^2$ because this function satisfies the invariance equation:

$$\varepsilon\frac{\partial x_1^2}{\partial x_1}x_1^2 = 2x_1x_1^2$$

and $\frac{\partial(x_1^2 - x_2)}{\partial x_2} = -1$.

Figure 5.2 demonstrates clearly the advantage of the modified control.

Example 13. We consider the control equation

$$\frac{d^3x}{dt^3} = u$$

with

$$S = \ddot{x} + \dot{x}^2 - x^2.$$

Setting

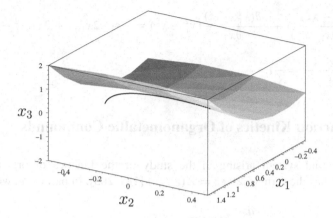

Fig. 5.2 Example 13. The hyperbolic paraboloid and the trajectory corresponding to the modified control law (*the solid line*)

$$x = x_1, \quad \dot{x} = x_2, \quad \ddot{x} = x_3,$$

we obtain the following control system

$$\dot{x}_1 = x_2, \quad \dot{x}_2 = x_3, \quad \dot{x}_3 = u,$$

with $S : x_3 - x_1^2 + x_2^2 = 0$ (hyperbolic paraboloid). In this case in accordance with (5.29) and (5.30)

$$f = \begin{pmatrix} x_2 \\ x_3 \\ 0 \end{pmatrix}, \quad B = \begin{pmatrix} 0 \\ 0 \\ 1 \end{pmatrix}, \quad \partial S / \partial x = G = (-2x_1, \ 2x_2, \ 1),$$

and $N = GBK = K$, where K is a scalar. Setting $K = 1$ we obtain $N = 1$. The modified control law is

$$u = -\frac{1}{\varepsilon} K \left[S(x) + \varepsilon N^{-1}(x) G(x) f(x) \right]$$

$$= -\varepsilon^{-1} \left(x_3 + x_1^2 - x_2^2 + \varepsilon(-2x_1 x_2 + 2x_2 x_3) \right)$$

and the corresponding guided system is

$$\dot{x}_1 = x_2, \quad \dot{x}_2 = x_3,$$

$$\varepsilon \dot{x}_3 = u = -x_3 + x_1^2 - x_2^2 + \varepsilon 2x_2(x_1 - x_3).$$

This last system has the attractive slow invariant manifold $x_3 = x_1^2 - x_2^2$ since this function satisfies the invariance equation

$$\varepsilon \frac{\partial (x_1^2 - x_2^2)}{\partial x_1} x_2 + \varepsilon \frac{\partial (x_1^2 - x_2^2)}{\partial x_2} (x_1^2 - x_2^2) = -\varepsilon(-2x_1x_2 + 2x_2(x_1^2 - x_2^2))$$

and $\frac{\partial(-x_3 + x_1^2 - x_2^2)}{\partial x_3} = -1.$

5.6 Reaction Kinetics of Organometallic Compounds

The differential system arising in the study of the kinetics of organometallic compounds has the form (5.2): $\varepsilon \dot{z} = Z(z, t, \varepsilon)$ [187, 205]. In particular, we have:

$$\varepsilon \frac{dz_1}{dt} = -\varepsilon a z_1 + b z_2, \tag{5.31}$$

$$\varepsilon \frac{dz_2}{dt} = \varepsilon a z_1 - b z_2 - c z_2 z_3 - z_2 z_4, \tag{5.32}$$

$$\varepsilon \frac{dz_3}{dt} = -c z_2 z_3, \tag{5.33}$$

$$\varepsilon \frac{dz_4}{dt} = -z_2 z_4. \tag{5.34}$$

The degenerate system $Z(z, 0) = 0$ is

$$0 = b z_2, \quad 0 = -b z_2 - c z_2 z_3 - z_2 z_4, \quad 0 = -c z_2 z_3, \quad 0 = -z_2 z_4$$

and it has a three-parameter family of solutions $z = \psi(v)$

$$z_1 = v_1, \quad z_3 = v_2, \quad z_4 = v_3, \quad z_2 = 0.$$

It is clear that the rank of ψ_v is equal to three, and the rank of matrix $A = Z_z(\psi, 0)$ is unity where

$$A = \begin{pmatrix} 0 & b & 0 & 0 \\ 0 & -b - cv_2 - v_3 & 0 & 0 \\ 0 & -cv_2 & 0 & 0 \\ 0 & -v_3 & 0 & 0 \end{pmatrix}$$

on noting $z_2 = 0$. The unique nonzero eigenvalue of this matrix is $-b - cv_2 - v_3$.

The system (5.31)–(5.34) may be reduced to the form (5.8), (5.9) by simply renaming the variables $z_1 = x_1, \ z_3 = x_2, \ z_4 = x_3, \ z_2 = y_2$:

$$\varepsilon \frac{dx_1}{dt} = -\varepsilon a x_1 + b y_2, \quad \varepsilon \frac{dx_2}{dt} = -c x_2 y_2, \quad \varepsilon \frac{dx_3}{dt} = -x_3 y_2, \tag{5.35}$$

$$\varepsilon \frac{dy_2}{dt} = \varepsilon a x_1 - b y_2 - c x_2 y_2 - x_3 y_2. \tag{5.36}$$

Setting $\varepsilon = 0$ in (5.36) we obtain that the function $y_2 = \varphi(x_1, x_2, x_3)$ is equal to zero identically. This implies (see Sect. 5.2) $y_2 = y_1 + \varphi = y_1$ and gives the possibility to use the change of variable $y_2 = y_1 = \varepsilon y$ to reduce the system (5.35) and (5.36) to

$$\frac{dx_1}{dt} = -ax_1 + by, \quad \frac{dx_2}{dt} = -cx_2 y, \quad \frac{dx_3}{dt} = -x_3 y, \tag{5.37}$$

$$\varepsilon \frac{dy}{dt} = ax_1 - by - cx_2 y - x_3 y \tag{5.38}$$

with three slow variables x_1, x_2, x_3 and one fast variable y. Note that the differential system (5.37) and (5.38) is linear with respect to the fast variable y, and this facilitates the construction of the asymptotic expansion for the three-dimensional attractive slow invariant manifold

$$y = h(x_1, x_2, x_3, \varepsilon) = h_0(x_1, x_2, x_3) + \varepsilon h_1(x_1, x_2, x_3) + O(\varepsilon^2).$$

The system (5.37), (5.38) has the form (2.2) and the coefficients of the asymptotic expansion may be found from the invariance equation (2.4) which becomes

$$\varepsilon \frac{\partial h(x_1, x_2, x_3, \varepsilon)}{\partial x_1} (-ax_1 + bh(x_1, x_2, x_3, \varepsilon))$$

$$+ \varepsilon \frac{\partial h(x_1, x_2, x_3, \varepsilon)}{\partial x_2} (-cx_2 h(x_1, x_2, x_3, \varepsilon))$$

$$+ \varepsilon \frac{\partial h(x_1, x_2, x_3, \varepsilon)}{\partial x_3} (-x_3 h(x_1, x_2, x_3, \varepsilon))$$

$$= ax_1 - (b + cx_2 + x_3)h(x_1, x_2, x_3, \varepsilon).$$

Setting $\varepsilon = 0$ in (5.38), we obtain

$$h_0(x_1, x_2, x_3) = ax_1/(b + cx_2 + x_3).$$

To calculate h_1 we equate the coefficients with the first power of ε in the invariance equation

$$\frac{\partial h_0(x_1, x_2, x_3)}{\partial x_1} (-ax_1 + bh_0(x_1, x_2, x_3)) + \frac{\partial h_0(x_1, x_2, x_3)}{\partial x_2} (-cx_2 h_0(x_1, x_2, x_3))$$

$$+ \frac{\partial h_0(x_1, x_2, x_3)}{\partial x_3} (-x_3 h_0(x_1, x_2, x_3)) = -(b + cx_2 + x_3)h_1(x_1, x_2, x_3).$$

This gives

$$h_1(x_1, x_2, x_3) = a^2 x_1 [(cx_2 + x_3)\Delta - c^2 x_1 x_2 - x_1 x_3]/\Delta^4,$$

where $\Delta = b + cx_2 + x_3$.

Returning to the initial variables z_1, z_2, z_3, z_4 and taking into account that $z_1 = x_1$, $z_3 = x_2$, $z_4 = x_3$, $z_2 = \varepsilon y$ it is found that the differential system (5.31)–(5.34) possesses the three-dimensional attractive slow invariant manifold

$$z_2 = \varepsilon a z_1 / \bar{\Delta} + \varepsilon^2 h_1(z_1, z_3, z_4),$$

where $\bar{\Delta} = b + cz_3 + z_4$, and $h_1(z_1, z_3, z_4) = a^2 z_1 [(cz_3 + z_4)\bar{\Delta} - c^2 z_1 z_3 - z_1 z_4]/\bar{\Delta}^4$. The flow on this manifold is described by equations

$$\frac{dz_1}{dt} = -az_1 + baz_1/\bar{\Delta} + \varepsilon b h_1 + O(\varepsilon^2), \tag{5.39}$$

$$\frac{dz_3}{dt} = -cz_3(az_1/\bar{\Delta} + \varepsilon b h_1) + O(\varepsilon^2), \tag{5.40}$$

$$\frac{dz_4}{dt} = -z_4(az_1/\bar{\Delta} + \varepsilon b h_1) + O(\varepsilon^2), \tag{5.41}$$

It should be noted in conclusion that the original differential system (5.31)–(5.34) has two first integrals

$$\frac{d}{dt}(z_1 + z_2 - z_3 - z_4) = 0, \quad \frac{d}{dt}(z_3 z_4^{-c}) = 0,$$

or

$$z_1 + z_2 - z_3 - z_4 = C_1, \quad z_3 z_4^{-c} = C_2$$

with arbitrary C_1, C_2. This allows us to further reduce the differential system under consideration. As result, it is possible to reduce the original differential system of the fourth order to a scalar differential equation without a singular perturbation

$$\frac{dz_4}{dt} = -z_4 \left[\frac{aC_2 z_4^c + az_4 + aC_1}{\bar{\Delta}} + \varepsilon a^2 (C_2 z_4^c + z_4 + C_1) \right.$$

$$\times \left(\frac{(z_4 + cC_2 z_4^c)\bar{\Delta} - c^2 C_2 z_4^c(z_4 + C_2 z_4^c + C_1) - z_4(z_4 + C_2 z_4^c + C_1)}{\bar{\Delta}^4} \right.$$

$$\left. \left. + \frac{1}{\bar{\Delta}^2} \right) \right] + O(\varepsilon^2),$$

where $\bar{\Delta} = b + cC_2 z_4^c + z_4, z_1 + z_2 - z_3 - z_4 = z_1(0) + z_2(0) - z_3(0) - z_4(0) = C_1, z_3 z_4^{-c} = z_3(0) z_4^{-c(0)} = C_2$.

Chapter 6
Reduction Methods for Chemical Systems

Abstract Many systems studied in chemical kinetics can be posed as a high order nonlinear differential system with slow and fast variables. This has given an impetus to the development of methods that reduce the order of the differential systems but retain a desired degree of accuracy. This research has led to a rapidly expanding volume of papers devoted to reduction methods. All these methods are connected with the integral manifold method in one way or another. These connections were clearly given by H. Kaper and T. Kaper in (Physica D 165:66–93, 2002), which also gives a good overview of reduction methods in chemical kinetics. In this chapter we will use results given previously in parallel with our interpretation of the connection between the two most often used reduction methods and demonstrate that both methods may be replaced successfully by regular procedures of approximation of slow integral manifolds which were described in Chap. 5.

6.1 Method of Intrinsic Manifolds

The method of intrinsic low-dimensional manifolds (ILDM Method) was proposed by Maas and Pope in [101] and developed in many later papers. This method, as applied to the system (2.2) in the form

$$\dot{x} = f(x, y, \varepsilon), \quad \dot{y} = \varepsilon^{-1} g(x, y, \varepsilon).$$

is based on a partition of the Jacobian matrix

$$J = J(x, y, \varepsilon) = \begin{pmatrix} \dfrac{\partial f}{\partial x} & \dfrac{\partial f}{\partial y} \\ \varepsilon^{-1} \dfrac{\partial g}{\partial x} & \varepsilon^{-1} \dfrac{\partial g}{\partial y} \end{pmatrix},$$

into fast and slow components at each point of (x-y)-space and a Schur decomposition [192] to generate bases for the corresponding fast and slow subspaces. This is a much more elaborate procedure than is necessary for the simplification of (2.2). The asymptotic method of slow integral manifolds in implicit form, discussed above in Sect. 4.1, was originally suggested by V. Sobolev several years before

© Springer International Publishing Switzerland 2014

E. Shchepakina et al., *Singular Perturbations*, Lecture Notes in Mathematics 2114, DOI 10.1007/978-3-319-09570-7_6

the publication of [101] (see, for example, [54], or [93]), and is simpler and more efficient. To illustrate this we restrict our attention to the system with scalar variables x and y.

Following [84], after calculations based on the Schur decomposition, it is possible represent the equation defined by the ILDM Method in the form

$$g_y g + \varepsilon g_x f - \varepsilon \lambda_s g = 0, \tag{6.1}$$

where λ_s is the eigenvalue

$$\lambda_s = \frac{1}{2}(\varepsilon^{-1} g_y + f_x) - \sqrt{\frac{1}{4}(\varepsilon^{-1} g_y + f_x)^2 - \varepsilon^{-1}(g_y f_x - f_y g_x)}$$

of the (2×2)-matrix J. When this result is compared with that of Eq. (4.6) in the autonomous case, viz.,

$$g_y g + \varepsilon g_x f = 0, \tag{6.2}$$

it is apparent that Eq. (6.1) includes the "unnecessary" term $-\varepsilon \lambda_s g$. As was shown in Sect. 4.1, Eq. (6.2) permits the calculation of the slow invariant manifold in the form

$$y = \phi(x) + \varepsilon h_1(x) + O(\varepsilon^2)$$

with an error of order $O(\varepsilon^2)$. In the paper [84] it is shown that, in general, the ILDM equation (6.1) gives the same error, and only in the case $\phi_{xx}(x) \equiv 0$ is the corresponding error $O(\varepsilon^3)$. However, it is more convenient to use the second order approximation equation in the form prior to (4.8) i.e.,

$$g_y g + \varepsilon g_x f + \varepsilon N_y g + \varepsilon^2 N_x f = 0,$$

on noting

$$N = g_y^{-1} g_x f,$$

rather than the ILDM equation. Moreover, Eq. (4.8) gives the error $O(\varepsilon^3)$ in the general case not only for planar systems but also in the case of vector variables x and y [84]. Note that the assumption $g_y < 0$ in the scalar case or the negativity of the real parts of all eigenvalues of the matrix g_y in the vector case, guarantees the manifolds are attractive and the corresponding implicit equations are solvable with respect to the fast variable by the implicit function theorem [146]. We illustrate the above with two examples.

Firstly we consider the system of two scalar differential equations which represent the model for the ignition of metal particles. Assuming an uniform temperature distribution in a particle, a constant particle size and the constant physical properties of both gas and particle, the known dimensionless model of the process has the following form [165]:

$$\frac{d\eta}{d\tau} = \varphi(\eta) \exp\left(\frac{\theta}{1 + \theta\beta}\right) = f(\eta, \theta),$$

$$\varepsilon\frac{d\theta}{d\tau} = \varphi(\eta) \exp\left(\frac{\theta}{1 + \beta\theta}\right) - \frac{\theta}{\kappa} = g(\eta, \theta),$$

(6.3)

with initial conditions

$$\eta(0) = 0, \quad \theta(0) = -\theta_i.$$

Here θ is the dimensionless temperature of a metal particle and

$$\theta = \frac{(T - T_0)E}{RT_0^2},$$

where T is the temperature of a metal particle and θ_i is a given initial temperature, T_0 is the gas temperature, E is the Arrhenius activation energy, R is the universal gas constant; η is the dimensionless related growth of the thickness of the oxide film and

$$\eta = \frac{\delta - \delta_{in}}{\delta},$$

where δ is the oxide film thickness, δ_{in} is the initial thickness of the film; τ is dimensionless time; the parameters β and ε reflect the temperature sensitivity and the exothermicity of the reaction; κ is a modified Semenov number related to convection heat transfer; $\varphi(\eta)$ is the kinetic function. Usually two forms of oxidation kinetics: $\varphi(\eta) = (\eta + 1)^{-n}$, $n = 1, 2$, corresponding to the cases of parabolic and cubic laws, respectively, are considered.

The initial temperature of a metal particle is either lower than the gas temperature or equal to it. In the first case, corresponding to cold particles, we have $\theta(0) = -\theta_i < 0$, and in the second, when the metal particles and the gas are heated simultaneously during a very short time, $\theta(0) = 0$.

The chemically relevant phase space Δ of the system (6.3) is defined by

$$\Delta := \{(\theta, \eta) \in \mathbb{R}^2 : \theta \geq -\theta_i, \eta \geq 0\}.$$

It should be noted that the system (6.3) is similar to the dimensionless model for the thermal explosion of a gas. But in the thermal explosion theory the kinetic function is usually $\varphi(\eta) = (1 - \eta)^n$, $n = 0, 1, 2$, and η reflects the depth of a gas conversion, while η is the dimensionless concentration of a gas.

In the case of very small particle size and significant initial thickness of the oxide film the parameter ε is small and, hence, the system (6.3) is singularly perturbed. Thus, it is possible to apply the implicit form for the construction of the slow invariant manifold in this case. Using Eq. (6.2) $g_\theta g + \varepsilon g_\eta f = 0$, where f and g are given in (6.3), we obtain the first order approximation to the slow invariant manifold

$$\left(\varphi(\eta) \frac{E(\theta)}{(1+\beta\theta)^2} - \frac{1}{\kappa} \right) \left(\varphi(\eta)E(\theta) - \frac{\theta}{\kappa} \right) + \varepsilon\varphi_\eta(\eta)\varphi(\eta)E^2(\theta) = 0,$$

where $E(\theta) = \exp\left(\frac{\theta}{1+\theta\beta} \right)$.

For our second example we consider again the model of the enzyme-substrate-exhibitor system from Sect. 3.5.2 viz., Eqs. (3.79)–(3.82), to construct an approximation to the slow invariant manifold. Here

$$f = \begin{pmatrix} -x_1 + (x_1 + K_s - L_s)y_1 + x_1 y_2 \\ \gamma[-x_2 + x_2 y_1 + (x_2 + K_i - L_i)]y_2 \end{pmatrix},$$

$$g = \begin{pmatrix} x_1 - (x_1 + K_s)y_1 - x_1 y_2 \\ \beta\gamma[x_2 - x_2 y_1 - (x_2 + K_i)y_2] \end{pmatrix},$$

and the first order approximation equation (6.2) gives the implicit equations with respect to y_1 y_2:

$$-(x_1 + K_s)[x_1 - (x_1 + K_s)y_1 - x_1 y_2] - x_1\beta\gamma[x_2 - x_2 y_1 - (x_2 + K_i)y_2]$$
$$+\varepsilon(1 - y_1 - y_2)[-x_1 + (x_1 + K_s - L_s)y_1 + x_1 y_2] = 0,$$
$$-\beta\gamma x_2[x_1 - (x_1 + K_s)y_1 - x_1 y_2] - \beta^2\gamma^2(x_2 + K_i)[x_2 - x_2 y_1 - (x_2 + K_i)y_2]$$
$$+\varepsilon\beta\gamma^2(1 - y_1 - y_2)[-x_2 + x_2 y_1 + (x_2 + K_i - L_i)y_2] = 0.$$

These give the same representations for the first order approximation of slow invariant manifold as in Sect. 3.5.2.

It should be noted that the use of implicit equations to describe slow integral manifolds can entail the occurrence of extraneous solutions. Thus, setting $\varepsilon = 0$ in (6.2), we obtain the equation $g_y g = 0$, which gives the extraneous solution $g_y = 0$ besides the equation of slow curve $g = 0$. After taking into account terms with the small parameter in a small neighborhood of the extraneous solution, the corresponding solution of (6.2) can be found and this solution will be also extraneous. It is clear that all solutions which do not belong to the ε order of the slow surface should be excluded from consideration.

6.2   Iterative Method

The iterative method was proposed by Fraser [46], and developed by Fraser and Roussel [145] for autonomous systems that are linear with respect to the fast variables in the case of scalar slow and fast variables.

Formally, the essence of the iterative method for the system

$$\dot{x} = \zeta(x, \varepsilon) + F(x, \varepsilon)y,$$
$$\varepsilon\dot{y} = \xi(x, \varepsilon) + G(x, \varepsilon)y,$$

(6.4)

i.e., (2.26) in the autonomous case with scalar variables, is as follows. It was found in Sect. 2.5 that for the autonomous system (6.4)

$$\varepsilon\frac{\partial h}{\partial x}(\zeta + Fh) = \xi + Gh$$

and then

$$h = \frac{-\xi + \varepsilon\zeta h_x}{G - \varepsilon F h_x}.$$

This representation is used to organize the iterative process by the algorithm

$$\varphi^{(0)} = \phi(x) = -\frac{\xi}{G}, \quad \varphi^{(k)} = \frac{-\xi + \varepsilon\zeta\varphi_x^{(k-1)}}{G - \varepsilon F\varphi_x^{(k-1)}}, \quad k = 1, 2, 3\dots,$$

where $\varphi^{(k)}$ is considered as an approximation to h. It can be shown that

$$h(x, \varepsilon) - \varphi^{(k)} = O(\varepsilon^{k+1}),$$

see [84].

Now we can extend the Fraser and Roussel [145] approach to the nonautonomous systems (2.26) in Sect. 2.5 with vector variables. We solve the equation

$$\varepsilon\frac{\partial h}{\partial t} + \varepsilon\frac{\partial h}{\partial x}(\zeta + Fh) = \xi + Gh,$$

for the vector function $h(x, t, \varepsilon)$ and obtain

$$h = (G - \varepsilon F h_x)^{-1}(-\xi + \varepsilon h_t + \varepsilon\zeta h_x).$$

As in the scalar case, this formula is the basis for the iterative procedure

$$\varphi^{(0)} = \phi(x, t) = -G^{-1}\xi, \quad \varphi^{(k)} = (G - \varepsilon F\varphi_x^{(k-1)})^{-1}(-\xi + \varepsilon\varphi_t^{(k-1)} + \varepsilon\zeta\varphi_x^{(k-1)}),$$

(6.5)

for the vector function $\varphi^{(k)}, k = 1, 2, 3 \ldots$. Thus, for example,

$$\varphi^{(1)} = (G - \varepsilon F \phi_x)^{-1}(-\xi + \varepsilon \phi_t + \varepsilon \zeta \phi_x),$$

and the asymptotic relationship

$$\|h(x, \varepsilon) - \varphi^{(k)}\| = O(\varepsilon^{k+1})$$

holds. We consider again the model from Sect. 3.5.2 viz., Eqs. (3.79)–(3.82) to construct an approximation to the slow invariant manifold.

On setting $\xi(x, 0) + G(x, 0)y = 0$ in (6.4), we obtain $y_1 = \bar{\phi}(x_1, x_2)$ and $y_2 = \bar{\bar{\phi}}(x_1, x_2)$. Then

$$\varphi^{(0)}(x) = \phi = \begin{pmatrix} \bar{\phi}(x_1, x_2) \\ \bar{\bar{\phi}}(x_1, x_2) \end{pmatrix} = \frac{1}{\Delta} \begin{pmatrix} K_i x_1 \\ K_s x_2 \end{pmatrix},$$

where $\Delta = K_s x_2 + K_i x_1 + K_s K_i$, and, consequently,

$$\phi_x = \begin{pmatrix} \dfrac{\partial \bar{\phi}}{\partial x_1} & \dfrac{\partial \bar{\phi}}{\partial x_2} \\ \dfrac{\partial \bar{\bar{\phi}}}{\partial x_1} & \dfrac{\partial \bar{\bar{\phi}}}{\partial x_2} \end{pmatrix} = \frac{K_i K_s}{\Delta^2} \begin{pmatrix} x_2 + K_i & -x_1 \\ -x_2 & x_1 + K_s \end{pmatrix}.$$

Now from (6.5) we obtain, on noting ϕ is independent of t,

$$\begin{aligned} \varphi^{(1)} &= (G - \varepsilon \phi_x F)^{-1}(-g + \varepsilon \phi_x \zeta) \\ &= \frac{1}{\Delta} \begin{pmatrix} K_i x_1 \\ K_s x_2 \end{pmatrix} + \varepsilon \frac{K_i K_s}{\beta \gamma \Delta^4} \begin{pmatrix} \beta \gamma (x_2 + K_i) P x_1 - x_1 x_2 Q \\ -\beta \gamma x_1 x_2 P + (x_1 + K_s) x_2 Q \end{pmatrix} + O(\varepsilon^2), \end{aligned}$$

where $P = (K_i L_s - \gamma K_s L_i)x_2 + K_i^2 L_s$, $Q = -(K_i L_s - \gamma K_s L_i)x_1 + \gamma K_s^2 L_i$. We note $\|h(x, \varepsilon) - \varphi^{(1)}\| = O(\varepsilon^2)$. And this agrees with the result in Sect. 3.5.2.

6.3 Extending the Iterative Method

Consider the differential system (2.1) and suppose that it possesses a slow integral manifold that can be found as an asymptotic expansion. Let ϕ be an isolated root of the degenerate equation $g(x, y, t, 0) = 0$. Using the change of variable $y = z + \phi(x, t)$ we obtain the system

$$\dot{x} = X(x, z, t, \varepsilon), \quad \varepsilon \dot{z} = B(x, t)z + Z(x, z, t, \varepsilon), \tag{6.6}$$

where

$$B(x,t) = \frac{\partial g}{\partial y}(x,\phi(x,t),t,0), \quad X(x,z,t,\varepsilon) = f(x,z+\phi(x,t),t,\varepsilon),$$

$$Z(x,z,t,\varepsilon) = g(x,z+\phi(x,t),t,\varepsilon) - \frac{\partial g}{\partial y}(x,\phi(x,t),t,0)z - \varepsilon\frac{\partial\phi}{\partial t}$$

$$-\varepsilon\frac{\partial\phi(x,t)}{\partial x}X(x,z,t,\varepsilon)).$$

The invariance equation (2.5) for this non-autonomous system takes the form

$$\varepsilon\frac{\partial\overline{h}}{\partial t} + \varepsilon\frac{\partial\overline{h}}{\partial x}X(x,\overline{h},t,\varepsilon)) = B(x,t)\overline{h} + Z(x,\overline{h},t,\varepsilon). \tag{6.7}$$

Due to the change of variable $y = z + \phi(x,t)$ the functions h and $\overline{h}$ that describe the slow integral manifolds for the systems (2.1) and (6.6), respectively, are connected by the equality $h = \overline{h} + \phi(x,t)$. By solving (6.7) for $\overline{h}$

$$\overline{h} = B(x,t)^{-1}(\varepsilon\frac{\partial\overline{h}}{\partial t} + \varepsilon\frac{\partial\overline{h}}{\partial x}X(x,\overline{h},t,\varepsilon) - Z(x,\overline{h},t,\varepsilon)),$$

this equation allows us to organize the iterative procedure in the following way, with the goal to obtain $\overline{h} = \varphi^{(k)} + O(\varepsilon^{k+1})$,

$$\varphi^{(0)} = 0,$$

$$\varphi^{(k)} = B(x,t)^{-1}(\varepsilon\frac{\partial\varphi^{(k-1)}}{\partial t} + \varepsilon\frac{\partial\varphi^{(k-1)}}{\partial x}X(x,\varphi^{(k-1)},t,\varepsilon)$$

$$-Z(x,\varphi^{(k-1)},t,\varepsilon)), \quad k = 1,2,\ldots.$$

Note, in conclusion, that the representation of the function $g(x,y,t,\varepsilon)$ in the form

$$g(x,y,t,\varepsilon) = g(x,y,t,0) + \varepsilon g_1(x,y,t,\varepsilon)$$

can be used to construct the following iterative procedure [84]:

$$g(x,\Phi^{(k)},t,0) = \varepsilon\frac{\partial\Phi^{(k-1)}}{\partial t} - \varepsilon\frac{\partial\Phi^{(k-1)}}{\partial x}f(x,\Phi^{(k-1)},t,\varepsilon)) - \varepsilon g_1(x,\Phi^{(k-1)},t,\varepsilon) \tag{6.8}$$

in the case when the function $\Phi^{(0)} = \phi$ can not be found in the explicit form. However, the use of that procedure is based on numerical differentiation. In such a situation the use of the slow integral manifold approximation in the implicit form is preferable, see Sect. 4.1.

Chapter 7
Specific Cases

Abstract The next two chapters consist of a contribution to advancing the geometrical approach to the investigation of singularly perturbed systems in cases when the main hypothesis is violated, i.e., when the real parts of some or all of the eigenvalues of the matrix of the linearized fast subsystem are no longer strictly negative. This means that the hypotheses of the Tikhonov's theorem are violated. This chapter is organized as follows. The first section is concerned with weakly attractive slow integral manifolds. The examples are borrowed from the theory of gyroscopic systems and flexible-joints manipulators. The next section is devoted to the application of repulsive slow invariant manifolds to thermal explosion problems. In the last section, the case when the slow integral manifold is conditionally stable is discussed and an optimal control problem is given as an application.

7.1 Weakly Attractive Integral Manifolds

In this section we consider the system (2.1) when the matrix $B = g_y(x, \phi(x,t), t, 0)$ has eigenvalues on the imaginary axis with nonvanishing imaginary parts. If the eigenvalues at $\varepsilon = 0$ are pure imaginary but after taking into account the perturbations of higher order they move to the complex left half-plane, then the system under consideration has stable slow integral manifolds. Some problems of the mechanics of gyroscopes and manipulators with high-frequency and weakly damped transient regimes are now discussed in this context. More results along this line can be found in [94, 131, 132, 134, 179, 197, 210]

7.1.1 Gyroscopic Systems

The general equations of gyroscopic systems on a fixed base may be represented in the Thomson–Tait form [12]:

$$\frac{dx}{dt} = y,$$

$$\varepsilon \frac{d}{dt}(Ay) = -(G + \varepsilon B)y + \varepsilon R + \varepsilon Q, \qquad (7.1)$$

© Springer International Publishing Switzerland 2014
E. Shchepakina et al., *Singular Perturbations*, Lecture Notes in Mathematics 2114,
DOI 10.1007/978-3-319-09570-7_7

where $R = R(x, y, t) = \frac{1}{2}\left[\frac{\partial(Ay)}{\partial x}\right]^T y$ is the result of differentiation of the quadratic forms components of the Routh function with respect to generalized coordinates x [102]. Here $x \in \mathbb{R}^n$, $A(t, x)$ is a symmetric positive definite matrix, $G(x, t)$ is a skew-symmetric matrix of gyroscopic forces, and $B(x, t)$ is a symmetric positive definite matrix of dissipative forces, $Q(x, t)$ is a vector of generalized forces and $\varepsilon = H^{-1}$ is a small positive parameter where H is the gyroscopic angular momentum. It is the tradition in the theory of gyroscopic and robotic systems to disregard non-dimensionalization (see, for example, [109, 190]), and we will follow that practice here.

The precessional equations take the form

$$(G + \varepsilon B)\frac{dx}{dt} = \varepsilon Q \tag{7.2}$$

which is obtained formally from (7.1) by setting $A = 0$.

Equation (7.2) are obtained from (7.1) by neglecting some of the terms multiplied by the small parameter ε. All roots of the characteristic equation

$$\det(G - \lambda I) = 0$$

are situated on the imaginary axis, since the matrix G is skew-symmetric, so that the inequality (2.6) is violated. To justify the use of the precessional equations we use the integral manifold method.

7.1.2   Precessional Motions

The gyroscopic system (7.1) has the slow integral manifold $y = \varepsilon h(x, t, \varepsilon)$, the motion on which is described by the differential system

$$\frac{dx}{dt} = \varepsilon h(x, t, \varepsilon). \tag{7.3}$$

Substituting $y = \varepsilon h(x, t, \varepsilon)$ in (7.1) and taking into account that $\varepsilon R(x, \varepsilon h, t) = \varepsilon \frac{1}{2}\left[\frac{\partial(A\varepsilon h)}{\partial x}\right]^T \varepsilon h = O(\varepsilon^3)$ and $\varepsilon \frac{d}{dt}(A\varepsilon h) = \varepsilon^2 \frac{\partial}{\partial t}Ah + O(\varepsilon^3)$ we obtain the equation

$$\varepsilon^2 \frac{\partial}{\partial t}(Ah) = -(G + \varepsilon B)\varepsilon h + \varepsilon Q + O(\varepsilon^3) \tag{7.4}$$

which allows us to represent (7.3) as

$$\frac{dx}{dt} = \varepsilon h(x, t, \varepsilon) = \varepsilon(G + \varepsilon B)^{-1}\left(Q - \varepsilon \frac{\partial}{\partial t}(Ah)\right) + O(\varepsilon^3). \tag{7.5}$$

Let $h(x,t,\varepsilon) = h_1(x,t) + \varepsilon h_2(x,t) + O(\varepsilon^2)$. Substituting this representation into (7.4) and equating powers of ε we obtain the relationships $0 = -Gh_1 + Q$ and $\frac{\partial}{\partial t}(Ah_1) = -Gh_2 - Bh_1$ which imply

$$h_1 = G^{-1}Q, \quad h_2 = -G^{-1}\left[Bh_1 + \frac{\partial(Ah_1)}{\partial t}\right]. \tag{7.6}$$

In the expressions for h_1, h_2, the matrices A, B, G and the function Q depend on x and t. Note that Eq. (7.3) describes slow precessional movements of the gyroscopic system (7.1).

Returning to the problem of the justification of precessional theory, we see that the r.h.s. of the precessional equations (7.2) in the form

$$\dot{x} = \varepsilon[G(x,t) + \varepsilon B(x,t)]^{-1}Q(x,t), \tag{7.7}$$

coincides with the r.h.s. of (7.4) with an accuracy of order $O(\varepsilon^2)$ in the case of a nonautonomous system, and with an accuracy of order $O(\varepsilon^3)$ in the case of an autonomous one. Taking into account that for the system under consideration fast nutation oscillations (nutation is a small and relatively rapid oscillation of the axis superimposed on the larger and much slower oscillation known as precession) are quenched, it may be inferred that the use of the precession equations is justified. More precisely, the Lyapunov reduction principle (see Chap. 2, formulae (2.7) and (2.8)) is valid for the slow integral manifold, but it is necessary to use the following inequalities

$$|\varphi_i(t,\varepsilon)| \leq N|y^0 - \varepsilon h(x^0,t_0,\varepsilon)| \exp[-\gamma(t - t_0)], \quad i = 1, 2, \tag{7.8}$$

instead of the inequalities (2.8) which contain the faster decaying function $\exp[-\gamma(t - t_0)/\varepsilon]$. In this section we present without proof some results from [12, 117] on investigations of integral manifolds in gyroscopic type systems.

7.1.3 Vertical Gyro with Radial Corrections

The equations of small oscillations of a gyroscopic system about the equilibrium position in dimensional variables have the form

$$A\ddot{x} + (HG + B)\dot{x} + Cx = 0,$$

where A is a symmetric positive-definite matrix of inertia, G is a skew-symmetric matrix of gyroscopic forces, B is a symmetric positive-definite matrix of dissipative forces, C is the matrix of potential and nonconservative forces, and H is the gyroscopic angular momentum. We let $\varepsilon = H^{-1}$ be a small positive parameter.

If G is a non-singular matrix, the characteristic roots of this linear autonomous system break down into groups of roots of order $O(\varepsilon)$ and order $O(1/\varepsilon)$. In those cases in which the roots of order $O(1/\varepsilon)$ lie in the left half-plane, we can set up a slow invariant manifold for which the reduction principle is valid, as in Chap. 2. The equations of motion on this manifold describe only precessional oscillations.

The corresponding precessional equations (setting $A = 0$), which can be considered as approximate equations on the integral manifolds, are

$$(HG + B)\dot{x} + Cx = 0.$$

Investigation of a vertical gyroscope with radial corrections leads to the equations

$$J\ddot{\alpha} - H\dot{\beta} + d\dot{\alpha} - k\beta = 0, \quad J\ddot{\beta} + H\dot{\alpha} + d\dot{\beta} + k\alpha = 0.$$

In these equations J is the equatorial mass moment of inertia of the gyroscope, $H > 0$ is its angular momentum, $d > 0$ is the coefficient of friction. Forces $-H\dot{\beta}$ and $H\dot{\alpha}$ are gyroscopic, while $-k\beta$ and $k\alpha$ are nonconservative forces which in the theory of gyroscopic systems are referred to as forces of radial corrections [109]. The roots of the corresponding characteristic equation, with $\alpha = \alpha_0 e^{\lambda t}$ and $\beta = \beta_0 e^{\lambda t}$,

$$\det \begin{vmatrix} J\lambda^2 + d\lambda & -H\lambda - k \\ H\lambda + k & J\lambda^2 + d\lambda \end{vmatrix} = 0$$

are

$$\lambda_{1,2} = -\nu \pm i\kappa, \quad \lambda_{3,4} = -\eta \pm i\gamma,$$

where

$$\nu = \left(d - \sqrt{(-m + \sqrt{n})/2} \right)/2J, \quad \kappa = \left(-H + \sqrt{(m + \sqrt{n})/2} \right)/2J,$$

$$\eta = \left(d + \sqrt{(-m + \sqrt{n})/2} \right)/2J, \quad \gamma = \left(H + \sqrt{(m + \sqrt{n})/2} \right)/2J,$$

$$n = (H^2 + d^2)^2 + 16Jk(Jk - Hd), \quad m = H^2 - d^2.$$

The orders of magnitude of the parameters in the governing equations give the following asymptotic representations

$$\nu = \varepsilon k + O(\varepsilon^3), \quad \kappa = -\varepsilon^2 kd + O(\varepsilon^3), \quad \eta = \frac{d}{J} - \varepsilon k + O(\varepsilon^2), \quad \gamma = \frac{1}{\varepsilon J} + O(1), \quad \varepsilon = 1/H.$$

The exact solution of the differential system under consideration may be represented in the form

$$\alpha(t) = \bar{\alpha}(t) + \varepsilon\varphi_1(t), \quad \beta(t) = \bar{\beta}(t) + \varepsilon\varphi_2(t),$$

where

$$\bar{\alpha}(t) = e^{-\nu t}(\xi\cos\kappa t - \zeta\sin\kappa t), \quad \bar{\beta}(t) = e^{-\nu t}(\xi\sin\kappa t + \zeta\cos\kappa t),$$

$$\varphi_1(t) = e^{-\eta t}(p\cos\gamma t + q\sin\gamma t), \quad \varphi_2(t) = e^{-\eta t}(-p\sin\gamma t + q\cos\gamma t),$$

and the coefficients ξ, ζ, p, q are defined by the initial values of $\alpha, \beta, \dot{\alpha}, \dot{\beta}$.

Thus, the solutions of the gyroscopic system are a sum of slowly varying functions $\bar{\alpha}(t), \bar{\beta}(t)$ (precessional movements), since $\nu = O(\varepsilon)$ and $\kappa = O(\varepsilon^2)$, and high-frequency and weakly damped motions $\varepsilon\varphi_1(t), \varepsilon\varphi_2(t)$ (nutational movements), since $\gamma = O(1/\varepsilon)$.

For this system the precessional equations take the form, by removing the second derivatives terms, (see (7.2))

$$H\dot{\beta} + k\beta - d\dot{\alpha} = 0, \quad H\dot{\alpha} + k\alpha + d\dot{\beta} = 0.$$

It is easy to see that, even in the case $d = 0$, the trivial solution of the precessional equations is asymptotically stable, whereas the trivial solution of the original equations is unstable, since $\eta < 0$ (see the asymptotic formulas above). This means that in the case $d = 0$ the use of precessional equations is inappropriate. The formula $\eta = \frac{d}{J} - \varepsilon k + O(\varepsilon^2)$ shows that the value of the damping factor for which asymptotic stability will prevail, i.e., $\eta > 0$, is

$$d > \varepsilon kJ = \frac{kJ}{H}.$$

We note that the angular momentum H of the gyroscope is large compared to kJ. Therefore, the lower limit for the damping factor is small, i.e., a small amount of damping renders the system stable. A more detailed analysis of this gyroscopic system is given in [109].

7.1.4 Heavy Gyroscope

Using either the general equations of motion or the Lagrange equations, the differential equations governing the motions of the axis of the heavy gyroscope in the Cardano suspension may be derived [91] as

$$A(\beta)\ddot{\alpha} + H\cos\beta \cdot \dot{\beta} + E\sin 2\beta \cdot \dot{\alpha} \cdot \dot{\beta} = -m_1\dot{\alpha},$$

$$B_0\ddot{\beta} - H\cos\beta \cdot \dot{\alpha} - \frac{1}{2}E\sin 2\beta \cdot \dot{\alpha}^2 = -m_2\dot{\beta} - Pl\cos\beta,$$

where $A(\beta) = (A + A_1)\cos^2\beta + C_1\sin^2\beta + A_2$, $B_0 = A + B_1$, $E = C_1 - A - A_1$.

Introduce the new time variable τ by $t = T\tau$ and the dimensionless parameters

$$\frac{A(\beta)}{B_0} = a(\beta), \quad \frac{E}{B_0} = e, \quad \frac{PlT}{B_0} = v,$$

$$\frac{m_1 T}{B_0} = b_1, \quad \frac{m_2 T}{B_0} = b_2, \quad B_0 T/H = \varepsilon,$$

and note that T can be chosen in such a way that $a(\beta)$, e, v, b_1, b_2 are parameters of order $O(1)$, and $\varepsilon \ll 1$.

Introducing the new dependent variables $\alpha_1 = \dot{\alpha}T$, $\beta_1 = \dot{\beta}T$ leads to the dimensionless equations

$$\frac{d\alpha}{d\tau} = \alpha_1, \quad \varepsilon a(\beta)\frac{d\alpha_1}{d\tau} + \cos\beta \cdot \beta_1 + \varepsilon e \sin 2\beta \cdot \alpha_1 \cdot \beta_1 = -\varepsilon b_1 \alpha_1,$$

$$\frac{d\beta}{d\tau} = \beta_1, \quad \varepsilon \frac{d\beta_1}{d\tau} - \cos\beta \cdot \alpha_1 - \varepsilon \frac{1}{2} e \sin 2\beta \cdot \alpha_1^2 = -\varepsilon b_2 \beta_1 - \varepsilon v \cos\beta.$$

Then with

$$x = \begin{pmatrix} \alpha \\ \beta \end{pmatrix}, \quad y = \begin{pmatrix} \alpha_1 \\ \beta_1 \end{pmatrix}, \quad A(x,t) = \begin{pmatrix} a(\beta) & 0 \\ 0 & 1 \end{pmatrix},$$

$$G(x,t) = \begin{pmatrix} 0 & \cos\beta \\ -\cos\beta & 0 \end{pmatrix}, \quad B(x,t) = \begin{pmatrix} b_1 & 0 \\ 0 & b_2 \end{pmatrix},$$

$$Q(x,t) = \begin{pmatrix} 0 \\ -v\cos\beta \end{pmatrix},$$

and with the new independent variable τ we have equations of the form (7.1). Note that the terms $\alpha_1 \beta_1$ and α_1^2 appear in $\frac{d}{d\tau}(Ay)$ and R respectively in (7.1):

$$\frac{d}{d\tau}(Ay) = \frac{d}{d\tau}\begin{pmatrix} a(\beta)\alpha_1 \\ \beta_1 \end{pmatrix} = \begin{pmatrix} a'(\beta)\beta_1\alpha_1 + a(\beta)\frac{d\alpha_1}{d\tau} \\ \frac{d\beta_1}{d\tau} \end{pmatrix}$$

$$= \begin{pmatrix} e \sin 2\beta \beta_1 \alpha_1 + a(\beta)\frac{d\alpha_1}{d\tau} \\ \frac{d\beta_1}{d\tau} \end{pmatrix}$$

and

$$R = \frac{1}{2}\left[\frac{\partial(Ay)}{\partial x}\right]^T y = \frac{1}{2}\left[\frac{\partial}{\partial x}\begin{pmatrix} a(\beta)\alpha_1 \\ \beta_1 \end{pmatrix}\right]^T \begin{pmatrix} \alpha_1 \\ \beta_1 \end{pmatrix}$$

$$= \frac{1}{2}\left[\begin{pmatrix} 0 & a'(\beta)\alpha_1 \\ 0 & 0 \end{pmatrix}\right]^T \begin{pmatrix} \alpha_1 \\ \beta_1 \end{pmatrix} = \frac{1}{2}e \sin 2\beta \begin{pmatrix} 0 \\ \alpha_1^2 \end{pmatrix},$$

since $a'(\beta) = e \sin 2\beta$, (see definitions of $A(\beta)$ and E). We used here the Jacobian matrix

$$\frac{\partial}{\partial x}\begin{pmatrix} f_1(x, y) \\ f_2(x, y) \end{pmatrix} = \begin{pmatrix} f_{1\alpha} & f_{1\beta} \\ f_{2\alpha} & f_{2\beta} \end{pmatrix}$$

for $f_1 = a(\beta)\alpha_1$ and $f_2 = \beta_1$, so that $f_{1\alpha} = 0$, $f_{1\beta} = a'(\beta)\alpha_1$, $f_{2\alpha} = 0$, $f_{2\beta} = 0$. From (7.6) we have

$$h_1 = G^{-1}Q = \begin{pmatrix} v \\ 0 \end{pmatrix}, \quad h_2 = -G^{-1}Bh_1 = \begin{pmatrix} 0 \\ -\frac{vb_1}{\cos\beta} \end{pmatrix},$$

since A and h_1 do not depend on t and, therefore, $\frac{\partial(Ah_1)}{\partial t} = 0$.

The motion on the slow invariant manifold is now described by the equations

$$\dot{\alpha} = \varepsilon v + O(\varepsilon^3),$$

$$\dot{\beta} = -\varepsilon^2 \frac{vb_1}{\cos\beta} + O(\varepsilon^3),$$

since $y = \dot{x} = \varepsilon h = \varepsilon h_1 + \varepsilon^2 h_2 + O(\varepsilon^3)$.

These relationships show that the motion of a heavy gyroscope is very close to a regular precession ($\dot{\alpha} = const$ and $\dot{\beta} = 0$) on bounded time intervals, but over times of order $O(1/\varepsilon^2)$ the angle β tends to the value $-\pi/2$, which is to say that the gyroscopic frames tend to the same plane. To show this, we rewrite the last differential equation for β, neglecting the terms of order $O(\varepsilon^3)$, in the form $\cos\beta \frac{d\beta}{d\tau} = -\varepsilon^2 vb_1$. After integration we obtain $\sin\beta = -\varepsilon^2 vb_1\tau + \sin\beta_0$, where β_0 is the initial value of β at $\tau = 0$. Note that we cannot use the apparatus of slow invariant manifolds near the point $\beta = -\pi/2$ since $\det G = 0$ at this point, but it is possible to show that $\beta \longrightarrow -\pi/2$ as $\tau \longrightarrow \infty$ using the Lyapunov functions technique, see [112].

7.1.5   Control of a One Rigid-Link Flexible-Joint Manipulator

Consider a simple model of a rigid-link flexible joint manipulator [190], where J_m is the motor inertia, J_1 is the link inertia, M is the link mass, l is the link length, c is the damping coefficient, k is the stiffness. The model is described by the equations:

$$J_1\ddot{q}_1 + Mgl\sin q_1 + c(\dot{q}_1 - \dot{q}_m) + k(q_1 - q_m) = 0,$$

$$J_m\ddot{q}_m - c(\dot{q}_1 - \dot{q}_m) - k(q_1 - q_m) = u.$$

Here q_1 is the link angle, q_m is the rotor angle, and u is the torque input which is the controller.

The control problem under consideration consists of a tracking problem in which it is desired that the link coordinate q_1 follows a time-varying smooth and bounded desired trajectory $q_d(t)$ so that $|q_d(t) - q_1(t)| \to 0$ as $t \to \infty$ [190].

If we rewrite the original system in the form

$$J_1 \ddot{q}_1 + J_m \ddot{q}_m + Mgl \sin q_1 = u,$$

$$\ddot{q}_1 - \ddot{q}_m + \frac{Mgl}{J_1} \sin q_1 + k \left(\frac{1}{J_1} + \frac{1}{J_m} \right) (q_1 - q_m) + c \left(\frac{1}{J_1} + \frac{1}{J_m} \right) (\dot{q}_1 - \dot{q}_m)$$

$$= -\frac{u}{J_m},$$

then the use of the small parameter $\varepsilon = 1/\sqrt{k}$ and new variables

$$x_1 = (J_1 q_1 + J_m q_m)/(J_1 + J_m), \quad x_2 = \dot{x}_1, \quad y_1 = q_1 - q_m, \quad y_2 = \varepsilon \dot{y}_1, \quad (7.9)$$

yields the system

$$\dot{x}_1 = x_2, \quad \dot{x}_2 = -\frac{Mgl}{J_1 + J_m} \sin \left(x_1 + \frac{J_m}{J_1 + J_m} y_1 \right) + \frac{u}{J_1 + J_m}, \quad (7.10)$$

$$\dot{\varepsilon} y_1 = y_2, \quad \varepsilon \dot{y}_2 = - \left(\frac{1}{J_1} + \frac{1}{J_m} \right) y_1 - \varepsilon c \left(\frac{1}{J_1} + \frac{1}{J_m} \right) y_2 \quad (7.11)$$

$$- \varepsilon^2 \frac{Mgl}{J_1} \sin \left(x_1 + \frac{J_m}{J_1 + J_m} y_1 \right) - \varepsilon^2 \frac{u}{J_m}.$$

Note that neglecting all terms of order $O(\varepsilon^2)$ in the r.h.s. of the last equation we obtain the independent subsystem

$$\dot{\varepsilon} y_1 = y_2,$$

$$\varepsilon \dot{y}_2 = - \left(\frac{1}{J_1} + \frac{1}{J_m} \right) y_1 - \varepsilon c \left(\frac{1}{J_1} + \frac{1}{J_m} \right) y_2,$$

solutions of which are characterized by high frequency $\approx \sqrt{(1/J_1 + 1/J_m)}/\varepsilon$ and relatively slow decay $c(1/J_1 + 1/J_m)/2$, since this differential system has the characteristic polynomial

$$\varepsilon^2 \lambda^2 + c \left(\frac{1}{J_1} + \frac{1}{J_m} \right) \lambda + \left(\frac{1}{J_1} + \frac{1}{J_m} \right)$$

which possesses complex zeros

$$\lambda_{1,2} = -\frac{c}{2} \left(\frac{1}{J_1} + \frac{1}{J_m} \right) \pm \frac{i}{\varepsilon} \sqrt{ \left(\frac{1}{J_1} + \frac{1}{J_m} \right) - \varepsilon^2 \frac{c^2}{4} \left(\frac{1}{J_1} + \frac{1}{J_m} \right)^2 }.$$

Since the real part of these numbers is negative, for the analysis of the manipulator model under consideration it is possible to use the slow invariant manifold noting that the reducibility principle holds for this manifold (the exact statement may be found in [117]). The terms of $O(\varepsilon^2)$ of the subsystem (7.11) lead us to conclude that the slow invariant manifold may be found in the form $y_1 = \varepsilon^2 Y + O(\varepsilon^3)$ and $y_2 = O(\varepsilon^3)$, where

$$Y = -\left[\frac{Mgl}{J_1}\sin(x_1) + \frac{u_0}{J_m}\right]\left(\frac{1}{J_1} + \frac{1}{J_m}\right)^{-1}. \tag{7.12}$$

Here we used the representation $u = u_0 + \varepsilon^2 u_1 + O(\varepsilon^3)$. Thus, the flow on this manifold is described by equations

$$\dot{x}_1 = x_2, \quad \dot{x}_2 = -\frac{Mgl}{J_1 + J_m}\sin\left(x_1 + \varepsilon^2\frac{J_m}{J_1 + J_m}Y\right) + \frac{u_0 + \varepsilon^2 u_1}{J_1 + J_m} + O(\varepsilon^3). \tag{7.13}$$

It is important to emphasize that due to (7.9) $q_1 = x_1 + \frac{J_m}{J_1 + J_m}y_1$, where $y_1 = \varepsilon^2 Y + O(\varepsilon^3)$, and on the slow invariant manifold we obtain the representation

$$q_1 = x_1 + \varepsilon^2\frac{J_m}{J_1 + J_m}Y + O(\varepsilon^3). \tag{7.14}$$

This allows us to rewrite the system (7.13) on the slow invariant manifold using the original variable q_1 instead x_1 in the form

$$\ddot{q}_1 - \varepsilon^2\frac{J_m}{J_1 + J_m}\ddot{Y} = -\frac{Mgl}{J_1 + J_m}\sin(q_1) + \frac{u_0 + \varepsilon^2 u_1}{J_1 + J_m} + O(\varepsilon^3). \tag{7.15}$$

The function $\ddot{Y}$ will be calculated below.

Let q_d be the desired trajectory, i.e., the goal of the controlled motion is $q_1 \to q_d$ as $t \to \infty$ [190]. Unlike [53, 190] we do not use a fast term added to the control input to make the fast dynamics asymptotically stable to guarantee the fast decay of fast variables y_1 and y_2, but we use the slow component of the control function u which is written as a sum $u_0 = (J_1 + J_m)u_d + Mgl\sin q_1$, where $u_d = \ddot{q}_d - a_1(x_1 - q_d) - a_2(\dot{x}_1 - \dot{q}_d)$ [190]. The goal of this control law is to obtain an equation with decaying solutions for the difference between q_1 and q_d.

Setting $\varepsilon = 0$, using (7.15) and the definitions of u_0 and u_d we obtain, to an accuracy of order $O(\varepsilon^2)$,

$$\ddot{q}_1 - \ddot{q}_d + a_2(\dot{q}_1 - \dot{q}_d) + a_1(q_1 - q_d) = 0$$

for the difference $q_1 - q_d$, since $q_1 = x_1 + O(\varepsilon^2)$ on the slow invariant manifold by (7.14). This differential equation allows us to choose the coefficients a_1 and a_2 in the control function u_d in such a way that the corresponding control function u

gives the possibility of realizing a desired motion. Let, for example [190], $M = 1$, $k = 100$, $l = 1$, $J_1 = 1$, $J_m = 1$, $g = 9.8$ and $c = 2$. Setting $a_1 = 3$, $a_2 = 4$ for the desired trajectory $q_d = \sin t$ we obtain the following control law for the original variables

$$u = (J_1 + J_m)u_d + Mgl \sin x_1 = (J_1 + J_m)u_d + Mgl \sin q_1 = 2u_d + 9.8 \sin q_1$$
$$= 2[-\sin t - 4(\dot{q}_1 - \cos t) - 3(q_1 - \sin t)] + 9.8 \sin q_1.$$

It is illustrated in Fig. 7.1, which contains the response of the controlled single link manipulator with given values of parameters, that the trajectory of q_1 tends to the desired trajectory $\sin t$ as t increases.

If it is necessary to take into account the terms of order $O(\varepsilon^2)$ we set $u_1 = -J_m \ddot{Y}$ to obtain the same equation $\ddot{q}_1 - \ddot{q}_d + a_2(\dot{q}_1 - \dot{q}_d) + a_1(q_1 - q_d) = 0$ from (7.15). To calculate Y we use (7.12) with $u_0 = (J_1 + J_m)u_d + Mgl \sin q_1$, where $u_d = \ddot{q}_d - a_1(x_1 - q_d) - a_2(\dot{x}_1 - \dot{q}_d)$, and obtain

$$Y = -Mgl \sin q_1 - J_1 u_d.$$

It is easy now to obtain $\dot{Y}$

$$\dot{Y} = -Mgl \cos q_1 \frac{dq_1}{dt} - J_1 \frac{du_d}{dt}$$
$$= -Mgl \cos q_1 \frac{dq_1}{dt} - J_1 \left[\frac{d^3 q_d}{dt^3} - a_1 \left(\frac{dq_1}{dt} - \frac{dq_d}{d} \right) - a_2 \left(\frac{d^2 q_1}{dt^2} - \frac{d^2 q_d}{dt^2} \right) \right]$$
$$= -Mgl \cos q_1 \frac{dq_1}{dt} - J_1 \left[\frac{d^3 q_d}{dt^3} + a_1 a_2(q_1 - q_d) + (a_2^2 - a_1) \left(\frac{dq_1}{dt} - \frac{dq_d}{dt} \right) \right],$$

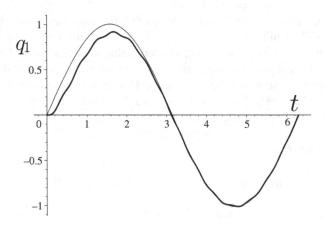

Fig. 7.1 The graph of $q_1(t)$ (*thin solid line*) and the desired trajectory $q_d = \sin t$ (*thick solid line*)

because $\ddot{q}_1 - \ddot{q}_d = -a_1(q_1 - q_d) - a_2(\dot{q}_1 - \dot{q}_d)$. Similarly we obtain

$$\ddot{Y} = Mgl \sin q_1 \left(\frac{dq_1}{dt}\right)^2 - Mgl \cos q_1 \frac{dq_1}{dt}$$

$$-J_1 \left[\frac{d^4 q_d}{dt^4} - a_1(a_2^2 - a_1)(q_1 - q_d) - a_2(a_2^2 - 2a_1)\left(\frac{dq_1}{dt} - \frac{dq_d}{dt}\right)\right],$$

and, finally, with $\varepsilon^2 = 1/k$,

$$u = (J_1 + J_m)u_d + Mgl \sin q_1 - J_m \ddot{Y}/k.$$

7.2 Unstable Manifolds

Consider the system (2.1) and suppose that hypothesis (I) holds, but inequality (2.6) is replaced by

$$Re\lambda_i(x, t) \geq 2\gamma > 0. \tag{7.16}$$

If in system (2.1) we use the new "reverse" time $t \to -t$, then we obtain a system that satisfies hypotheses (I) and (II). Consequently (2.1) has the slow integral manifold $y = h(x, t, \varepsilon)$, and for this manifold all propositions from the previous section are true, with the exception of stability.

But this manifold is stable (and the Lyapunov Reduction Principle applies, see Sect. 2.2) with respect to $t \to -\infty$. This means that for increasing t, the trajectories of solutions with initial points near the slow integral manifold move away from this manifold very rapidly.

Now we will demonstrate how the unstable slow integral manifold can be useful for modelling the critical regime which separates an explosive regime from a non-explosive one. We use the example of a classical combustion model. Following [107] it is possible to say that a one-dimensional unstable slow invariant manifold plays the role of a *watershed line* in this problem.

Thermal explosion occurs when chemical reactions produce heat too rapidly for a stable balance between heat production and heat loss. The exothermic oxidation reaction is usually modelled as a single step reaction obeying an Arrhenius temperature dependence. The first model for the self-ignition was constructed by Semenov in 1928 (see, for example [3, 152]). The basic idea of the model was a competition between heat production in the reactant vessel (due to an exothermic reaction) and heat losses on the vessel's surface. Heat losses were assumed proportional to the temperature excess over the ambient temperature (Newtonian cooling). The main assumption was that there is no reactant conversion during the fast highly exothermic reaction. This assumption implies the absence of the energy conservation law in the model, and gave the possibility of constructing an extremely simple and attractive mathematical model. Spatial uniformity of the temperature was also assumed so that the governing equation was one first-order ordinary differential equation for the temperature changes:

$$c\rho V \frac{dT}{dt} = QV\left(-\frac{dC}{dt}\right) - \chi S(T - T_0),$$

where

$$-\frac{dC}{dt} = \Psi(C)A\exp\left(-\frac{E}{RT}\right),$$

and Ψ expresses the dependence of reaction rate on reactant concentration. Here Q is an exothermicity per mole reactant; C and C_0 are a reactant concentration and its initial value; A is constant which is known as a pre-exponential rate factor; c is specific heat capacity; ρ is reactant density; χ is the heat-transfer coefficient; E is the Arrhenius activation energy; R is the universal gas constant; V is the reactant vessel volume; S is the surface area of the reactant vessel; t is a time variable; T is absolute temperature; T_0 is ambient temperature. The initial temperature is assumed to be equal to the ambient temperature T_0.

Dimensionless variables τ, η, θ are introduced by

$$\tau = tC_0^{n-1}A\exp\left(-\frac{E}{RT_0}\right), \quad \eta = 1 - C/C_0, \quad \theta = \frac{E}{RT_0}(T - T_0),$$

(n is the order of the chemical reaction) and we obtain the classical model of thermal explosion with reactant consumption [65, 219]:

$$\varepsilon\frac{d\theta}{d\tau} = \Psi(\eta)\exp\left(\theta/\left(1 + \beta\theta\right)\right) - \alpha\theta, \tag{7.17}$$

$$\frac{d\eta}{d\tau} = \Psi(\eta)\exp\left(\theta/\left(1 + \beta\theta\right)\right), \tag{7.18}$$

$$\eta(0) = \eta_0/\left(1 + \eta_0\right) = \bar{\eta}_0, \quad \theta(0) = 0.$$

Here the parameter η_0 is a kinetic parameter (the ratio of the initial reaction rate to an autocatalytic constant), where the small dimensionless parameters

$$\beta = \frac{RT_0}{E} \quad \text{and} \quad \varepsilon = \frac{c\rho}{QC_0}\frac{E}{RT_0^2}$$

characterize the physical properties of the gas mixture, and

$$\alpha = \frac{\chi S}{VQC_0^nA}\frac{RT_0^2}{E}\exp\left(\frac{E}{RT_0}\right) \tag{7.19}$$

is the dimensionless heat loss parameter.

The following cases are examined: $\Psi(\eta) = 1 - \eta$ ($\eta_0 = 0$) (first-order reaction) and $\Psi(\eta) = \eta(1-\eta)$ (autocatalytic reaction). The system (7.17), (7.18) is singularly perturbed. According to the standard approach to such systems the limiting case $\varepsilon \to 0$ is examined, and discontinuous solutions of the reduced system are analyzed. This makes it possible to determine some critical values of initial conditions, which provide a jump transition from the slow regime to the explosive ones. The study of transitional regimes requires the application of higher approximations in the asymptotic analysis of the systems of the type given in Eqs. (7.17), (7.18). The integral manifold technique is applied to the qualitative analysis of critical and transitional regimes for both types of chemical reaction: the case of the first-order reaction is considered below and in the Example 11, while the case of the autocatalytic reaction is examined in Sect. 8.5.3.

For the first-order reaction, when $\Psi(\eta) = 1 - \eta$ and the dimensionless concentration $\bar{\eta} = 1 - \eta$ replaces η, the system (7.17), (7.18) is

$$\varepsilon \frac{d\theta}{d\tau} = \bar{\eta} \exp\left(\theta/\left(1 + \beta\theta\right)\right) - \alpha\theta, \tag{7.20}$$

$$\frac{d\bar{\eta}}{d\tau} = -\bar{\eta} \exp\left(\theta/\left(1 + \beta\theta\right)\right). \tag{7.21}$$

The initial conditions are

$$\bar{\eta}(0) = 1, \quad \theta(0) = 0. \tag{7.22}$$

From the expression (7.19) one can see that the parameter α characterizes the initial physical state of the chemical system. Depending on its value the chemical reaction either changes to a slow regime with decay of the reaction, or into a regime of self-acceleration which leads to an explosion. For some value of α (we call it critical) the reaction is maintained and gives rise to a sharp transition from slow reactions to explosive ones. The transition region from slow regimes to explosive ones exists due to the continuous dependence of the system (7.20), (7.21) on the parameter α. To find the critical value of the parameter α, we may use special asymptotic formulae given in [111]. The equation

$$\bar{\eta} \exp\left(\theta/\left(1 + \beta\theta\right)\right) - \alpha\theta = 0 \tag{7.23}$$

gives the slow curve S of the system (7.20), (7.21). The curve S has two jump points (A_1 and A_2), see Fig. 7.2, where the slope is zero, given by the equation

$$\frac{\partial}{\partial\theta} \left(\bar{\eta} \exp\left(\theta/\left(1 + \beta\theta\right)\right) - \alpha\theta\right) = 0,$$

i.e.

$$\bar{\eta} \left(1 + \beta\theta\right)^{-2} \exp\left(\theta/\left(1 + \beta\theta\right)\right) - \alpha = 0. \tag{7.24}$$

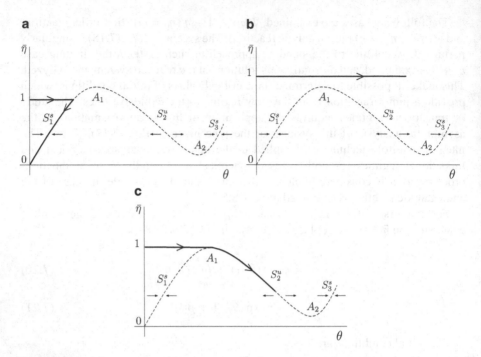

Fig. 7.2 The slow curve (the *dashed line*) and the trajectory (the *solid line*) of the system (7.20), (7.21) in the limit case ($\varepsilon = 0$): **(a)** in the case of a slow combustion regime, **(b)** in the case of the thermal explosion when θ becomes large and the trajectory does not lie on the slow integral manifold till S_3^s, **(c)** in the case of a critical regime

The equation

$$\theta - (1 + \beta\theta)^2 = 0,$$

which is implied from (7.23) and (7.24), determines the coordinates θ of the jump points. The point A_1 has the coordinates $\theta = \theta_1 = 1 + 2\beta + 0(\beta^2)$ and $\bar{\eta} = \eta_1 = \alpha(1 + \beta)/e + 0(\beta^2)$ from (7.24), while the point A_2 corresponds to a large value of the temperature $\theta = O(\beta^{-2})$ as $\beta \to 0$.

The jump points divide the slow curve into three parts S_1^s, S_2^u, S_3^s (see Fig. 7.2) which are zeroth order approximations for the corresponding slow integral manifolds $S_{1,\varepsilon}^s$, $S_{2,\varepsilon}^u$ and $S_{3,\varepsilon}^s$. Manifolds $S_{1,\varepsilon}^s$ and $S_{3,\varepsilon}^s$ are stable and $S_{2,\varepsilon}^u$ is unstable. Each manifold $S_{1,\varepsilon}^s$, $S_{2,\varepsilon}^u$ and $S_{3,\varepsilon}^s$ is at the same time part of some trajectory of the system (7.20), (7.21).

For some values of α, trajectories of equations (7.20)–(7.22) move along the manifold $S_{2,\varepsilon}^u$, sooner or later either falling into an explosive regime (if the trajectories jump from $S_{2,\varepsilon}^u$ toward $S_{3,\varepsilon}^s$), or rapidly passing into a slow regime (if the trajectories jump from $S_{2,\varepsilon}^u$ toward $S_{1,\varepsilon}^s$), see Fig. 7.2c. The value of $\alpha = \alpha_2$, at which the trajectory $\mathcal{T}_2$ of (7.20)–(7.22) contains the unstable manifold $S_{2,\varepsilon}^u$ (see

Fig. 7.3), is said to be critical, i.e. this regime is not a slow combustion regime, since θ achieves a high value, and is not explosive, as the temperature increases at the tempo of the slow variable as this is the slow manifold. The value $\alpha = \alpha_1$, for which the trajectory $\mathcal{T}_1$, corresponding to the limit case for all trajectories, as it passes through A_1, shown in Fig. 7.2a, being the longest one, contains the manifold $S_{1,\varepsilon}^s$ (see Fig. 7.4), and is called the slow critical value. The trajectory $\mathcal{T}_3$ contains the stable manifold $S_{3,\varepsilon}^s$, see Fig. 7.4, and does not determine any critical regime, since it does not begin from the initial point P. We note that any trajectory of the system starting at the point $\bar{\eta} = 1$, $\theta = 0$, i.e. the initial point, runs to the left of $\mathcal{T}_3$ because different trajectories cannot intersect, ending up at $\bar{\eta} = \theta = 0$. Since the analysis of the trajectories near the stable slow manifolds is not our goal we will concentrate our attention on the transition region.

The value of $\alpha = \alpha_1$ gives the critical trajectory $\mathcal{T}_1$, see Fig. 7.4. It separates the transition region (see Fig. 7.5) from slow regimes (see Fig. 7.2a) which are characterized by a slowdown of the reaction with small degrees of conversion and heating up is limited from above by $\theta < \theta_1$ at A_1.

The region of slow transitional trajectories corresponds to the interval (α_2, α_1). These trajectories are characterized by a comparatively rapid (but not explosive) flow of the reaction till the essential degree of conversion takes place and then a jump slow-down (horizontal line) and a transition to the slow flow of the reaction to near the origin, see Fig. 7.5. For $\alpha \approx \alpha_1$ the slow transitional trajectories of the

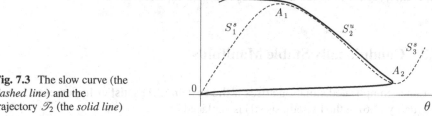

Fig. 7.3 The slow curve (the *dashed line*) and the trajectory $\mathcal{T}_2$ (the *solid line*)

Fig. 7.4 The slow curve (the *dashed line*) and the trajectories $\mathcal{T}_1$ and $\mathcal{T}_3$ (the *solid line*)

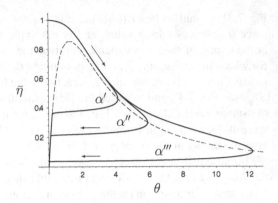

Fig. 7.5 The slow curve and
the trajectories
of (7.20)–(7.22) for $\varepsilon = 0.01$,
$\beta = 0.1, \alpha' = 2.08039$,
$\alpha'' = 2.0803865$,
$\alpha''' = 2.080386$
$(\alpha_2 < \alpha''' < \alpha'' < \alpha' < \alpha_1)$,
where the arrows indicate
increasing time. When
$\alpha > \alpha_1$ we have the slow
regime and when $\alpha < \alpha_2$ we
have the explosion

system are close to the trajectory $\mathscr{T}_1$. For the smaller values of the parameter α the
trajectories of the system remain for a longer time on the unstable slow manifold
$S_{2,\varepsilon}^u$ and the trajectory $\mathscr{T}_2$ is the longest of these. For $\alpha < \alpha_2$ the trajectories are
explosive where θ achieves a high value, see Fig. 7.2b.

The critical value $\alpha = \alpha_2$ corresponding to the trajectory $\mathscr{T}_2$ may be obtained by
means of the asymptotic expansion

$$\alpha_2 = e(1 - \beta)\left[1 - \Omega_0\sqrt[3]{2}\left(1 + \frac{7}{3}\beta\right)\varepsilon^{2/3} + \frac{4}{9}(1 + 6\beta)\varepsilon \ln\frac{1}{\varepsilon}\right] + O(\varepsilon + \beta^2),$$

where $\Omega_0 = 2.338107$, see [111] for the details.

The crucial result is that the unstable slow manifold may be used to construct the
separating regime between the safe regimes and explosive ones.

7.3   Conditionally Stable Manifolds

A rather complicated situation arises if the system (2.1) satisfies hypothesis (I) but
inequality (2.6) in the hypothesis (II) is replaced by

$$Re\lambda_i(x,t) \geq 2\gamma_1 > 0, i = 1,\ldots,n_1, \tag{7.25}$$

$$Re\lambda_i(x,t) \leq -2\gamma_2 < 0, i = n_1 + 1,\ldots,n. \tag{7.26}$$

The slow integral manifold $y = h(x,t,\varepsilon)$ is then conditionally stable, i.e. in the
space $\mathbb{R}^m \times \mathbb{R}^n$ there exists an n_2-dimensional manifold $(n_2 = n - n_1)$, which has the
following property: all trajectories, with initial points that belong to this manifold,
tend to the slow integral manifold as $t \to \infty$. Besides, in $\mathbb{R}^m \times \mathbb{R}^n$ there exists an
n_1-dimensional manifold such that all trajectories, with initial points that belong to
this manifold, tend to the slow integral manifold as $t \to -\infty$. Such a situation is
typical of optimal control problems [170].

As an example consider the problem of the minimization of the functional

$$I_\varepsilon = \frac{1}{2} \int_0^1 [x_1^2(t) + x_2^2(t) + u^2(t)]dt \qquad (7.27)$$

under the restrictions

$$\dot{x}_1 = x_2, \quad x_1(0) = x_{01}, \quad \varepsilon\dot{x}_2 = w(x_1) + u, \quad x_2(0) = x_{02}, \qquad (7.28)$$

where u is a control function. A more general situation is considered in the book [92] where the necessary optimality conditions are derived in detail using the Lagrangian and Hamiltonian formulations. We formulate a necessary optimality condition using the Hamiltonian [92]

$$H = \frac{1}{2}(x_1^2 + x_2^2 + u^2) + px_2 + q(w(x_1) + u),$$

where p and q are the multipliers (costates or adjoint variables) associated with x_1 and x_2 respectively. The Hamiltonian necessary conditions are (see, for example, [92])

$$0 = \frac{\partial H}{\partial u}; \quad \dot{x}_1 = \frac{\partial H}{\partial p}, \quad \varepsilon\dot{x}_2 = \frac{\partial H}{\partial q}; \quad \dot{p} = -\frac{\partial H}{\partial x_1}; \quad \varepsilon\dot{q} = -\frac{\partial H}{\partial x_2}$$

with boundary value conditions

$$x_1(0) = x_{01}, \quad x_2(0) = x_{02}, \quad p(1) = 0, \quad q(1) = 0. \qquad (7.29)$$

Thus, we have for u the condition $\frac{\partial H}{\partial u} = u + q = 0$, i.e.,

$$u = -q \qquad (7.30)$$

and the boundary value problem can be represented as

$$\dot{x}_1 = \frac{\partial H}{\partial p} = x_2, \quad \varepsilon\dot{x}_2 = \frac{\partial H}{\partial q} = w(x_1) - q, \qquad (7.31)$$

$$\dot{p} = -\frac{\partial H}{\partial x_1} = -x_1 - w'(x_1)q, \quad \varepsilon\dot{q} = -\frac{\partial H}{\partial x_2} = -p - x_2, \qquad (7.32)$$

$$x_1(0) = x_{01}, \quad x_2(0) = x_{02}, \quad p(1) = 0, \quad q(1) = 0$$

where $w'(x_1) = \frac{\partial w}{\partial x_1}$.

Note that this system is linear with respect to the fast variables x_2 and q.

Setting

$$x = \begin{pmatrix} x_1 \\ p \end{pmatrix}, \quad y = \begin{pmatrix} x_2 \\ q \end{pmatrix},$$

and

$$\zeta = \begin{pmatrix} 0 \\ -x_1 \end{pmatrix}, \quad \xi = \begin{pmatrix} w(x_1) \\ -p \end{pmatrix}, \quad F = \begin{pmatrix} 1 & 0 \\ 0 & -w'(x_1) \end{pmatrix}, \quad G = \begin{pmatrix} 0 & -1 \\ -1 & 0 \end{pmatrix},$$

we represent the system (7.31) and (7.32) in the form (2.26) (see Sect. 2.5):

$$\dot{x} = \zeta(x) + F(x)y, \quad \varepsilon\dot{y} = \xi(x) + Gy. \tag{7.33}$$

The eigenvalues of the constant matrix G are ± 1 and this means that the slow invariant manifold is conditionally stable. Using the formulas (2.27) we obtain the first order approximation of this manifold in the form

$$y = \phi + \varepsilon h_1 = -G^{-1}\xi + \varepsilon G^{-1}\frac{\partial\phi}{\partial x}(\zeta + F\phi) = \begin{pmatrix} -p \\ w \end{pmatrix} + \varepsilon\begin{pmatrix} w'p \\ -x_1 - w'w \end{pmatrix} \tag{7.34}$$

on noting

$$\frac{\partial\phi}{\partial x} = \begin{pmatrix} 0 & -1 \\ w' & 0 \end{pmatrix}.$$

Then the motion on the slow invariant manifold is described by

$$\dot{x} = \zeta + F\phi + \varepsilon F h_1 + O(\varepsilon^2) = \begin{pmatrix} -(1 - \varepsilon w')p \\ -(x_1 + w'w)(1 - \varepsilon w') \end{pmatrix} + O(\varepsilon^2). \tag{7.35}$$

The use of new the variable $z = y - h(x, \varepsilon)$ in (7.33) leads to the equation $\varepsilon\dot{z} = \varepsilon\dot{y} - \varepsilon\frac{\partial h}{\partial x}\dot{x} = \xi + Gh + Gz - \varepsilon\frac{\partial h}{\partial x}(\zeta + Fh + Fz) = Gz - \varepsilon\frac{\partial h}{\partial x}Fz$ due to the invariance equation $\varepsilon\frac{\partial h}{\partial x}(\zeta + Fh) = \xi + Gh$ and $\dot{x} = \zeta + Fh + Fz$. As the result, we obtain the system

$$\dot{x} = \zeta + Fh + Fz, \tag{7.36}$$

$$\varepsilon\dot{z} = (G - \varepsilon\frac{\partial h}{\partial x}F)z = (G + O(\varepsilon))z, \tag{7.37}$$

where the second equation is approximately independent of x. To exclude z from Eq. (7.36) we need to use the change of variables $x = v + \varepsilon P(v, z, \varepsilon)$, where the function P describes the so called fast invariant manifold [117, 170]. However for our purposes it is enough to note that the change of variables $x = v + \varepsilon F G^{-1}z$ in (7.36) where

$$v = \begin{pmatrix} v_1 \\ p_1 \end{pmatrix}, \quad z = \begin{pmatrix} z_1 \\ q_1 \end{pmatrix} \tag{7.38}$$

leads to the equation

$$\dot{v} = \zeta(v) + F(v)(\phi(v) + \varepsilon h_1(v)) + O(\varepsilon^2) + O(\varepsilon \|z\|), \tag{7.39}$$

on using (7.37) and recalling $h = \phi + \varepsilon h_1 + +O(\varepsilon^2)$. Here $\|z\|$ characterizes the distance between y and the slow invariant manifold $y = h(x, \varepsilon)$ and (7.39) describes the motion on this manifold. Formally, (7.39) is obtained from (7.35) by setting $x = v$ (or in scalar form, $x_1 = v_1$, $p = p_1$). If we neglect terms of order $O(\varepsilon^2)$, the change of variables $y = z + h(x, \varepsilon)$ written in scalar form is

$$x_2 = z_1 - p + \varepsilon w'(x_1)p, \quad q = q_1 + w(x_1) - \varepsilon(x_1 + w'(x_1)w(x_1)), \tag{7.40}$$

on using (7.38) for z and (7.34) for h and noting $h = \phi + \varepsilon h_1$. The change of variables $x = v + \varepsilon FG^{-1}z$ written in scalar form is

$$x_1 = v_1 - \varepsilon q_1, \quad p = p_1 + \varepsilon w'(v_1)z_1, \tag{7.41}$$

on using (7.38) for v and z.

Noting that (7.39) is obtained from (7.35) by setting $x = v$ (or in scalar form, $x_1 = v_1$, $p = p_1$), we can to rewrite (7.39) in scalar form to get

$$\dot{v}_1 = -(1 - \varepsilon w')p_1, \quad \dot{p}_1 = -(v_1 + w'w)(1 - \varepsilon w').$$

Terms of order $O(\varepsilon^2) + O(\varepsilon \|z\|)$ in these equations are neglected. Using the additional transformation $z_1 = z_2 + q_2$, $q_1 = z_2 - q_2$ (or $z_2 = (z_1 + q_1)/2$, $q_2 = (z_1 - q_1)/2$), which diagonalizes the matrix G, we obtain from (7.37) $(\varepsilon \dot{z} = (G + O(\varepsilon))z$, i.e. $\varepsilon \dot{z}_1 = -q_1$, $\varepsilon \dot{q}_1 = -z_1)$

$$\varepsilon \dot{z}_2 = -z_2, \quad \varepsilon \dot{q}_2 = q_2,$$

where $w = w(v_1)$ and $w' = w'(v_1)$, since $\varepsilon \dot{z}_2 = (\varepsilon \dot{z}_1 + \varepsilon \dot{q}_1)/2 = (-q_1 - z_1)/2 = -z_2$ and $\varepsilon \dot{q}_2 = (\varepsilon \dot{z}_1 - \varepsilon \dot{q}_1)/2 = (-q_1 + z_1)/2 = q_2$. Terms of order $O(\varepsilon)$ in the last two equations are neglected. The main purpose of introducing z_2 and q_2 is to use the asymptotic formula $\exp \alpha/\varepsilon = o(\varepsilon^k)$, where α is any positive number and k is any positive integer, and we obtain the following representation for $z_2(t, \varepsilon)$ and $q_2(t, \varepsilon)$

$$z_2(1, \varepsilon) \simeq 0, \quad q_2(0, \varepsilon) \simeq 0, \tag{7.42}$$

since $z_2 = Ae^{t/\varepsilon}$ and $q_2 = Be^{(t-1)/\varepsilon}$ where A and B are arbitrary constants. The resulting change of variables for x_1 and p in scalar form is, from (7.41) with $z_1 = z_2 + q_2$, $q_1 = z_2 - q_2$,

$$x_1 = v_1 - \varepsilon q_1 = v_1 - \varepsilon(z_2 - q_2), \quad p = p_1 + \varepsilon w'(v_1)z_1 = p_1 + \varepsilon w'(v_1)(z_2 + q_2)$$
$$(7.43)$$

and for x_2, q from (7.40):

$$x_2 = z_2 + q_2 - p + \varepsilon w'(x_1)p, \quad q = z_2 - q_2 + w(x_1) - \varepsilon(x_1 + w'(x_1)w(x_1)). \quad (7.44)$$

Using the initial values in the first of (7.43) gives

$$x_{01} = x_1(0) = v_1(0) - \varepsilon(z_2(0) - q_2(0)) = v_1(0) - \varepsilon z_2(0),$$

on noting the second of (7.42). The second of (7.43) and the first of (7.42) gives

$$0 = p(1) = p_1(1) + \varepsilon w'(v_1(1))(z_2(1) + q_2(1)) = p_1(1) + \varepsilon w'(v_1(1))q_2(1).$$

Similarly, (7.44) gives

$$x_{02} = x_2(0) = z_2(0) + q_2(0) - p(0) + \varepsilon w'(x_1(0))p(0)$$
$$= z_2(0) - p(0) + \varepsilon w'(x_{01})p(0),$$

and

$$0 = q(1) = z_2(1) - q_2(1) + w(x_1(1)) - \varepsilon(x_1(1) + w'(x_1(1))w(x_1(1)))$$
$$= -q_2(1) + w(x_1(1)) - \varepsilon(x_1(1) + w'(x_1(1))w(x_1(1))).$$

The last two equalities are rewritten in the form

$$x_{02} = z_2(0) - p_1(0) + O(\varepsilon), \quad 0 = -q_2(1) + w(v_1(1)) + O(\varepsilon).$$

It is now a straightforward exercise from the above equations to check that the new variables v_1 and p_1, z_2 and q_2 satisfy the following initial value and boundary value conditions

$$z_2(0) = x_{02} + p_1(0), \qquad q_2(1) = w(v_1(1))$$
$$v_1(0) = x_{01} + \varepsilon z_2(0) \Rightarrow v_1(0) - \varepsilon p_1(0) = x_{01} + \varepsilon x_{20},$$
$$p_1(1) + \varepsilon w'(v_1(1))w(v_1(1)) = 0,$$

on neglecting terms of order $O(\varepsilon)$ in the expressions for $z_2(0)$ and $q_2(1)$. As a result we obtain the independent second order boundary value problem

$$\dot{v}_1 = -(1 - \varepsilon w')p_1, \quad \dot{p}_1 = -(v_1 + w'w)(1 - \varepsilon w'),$$

with the boundary conditions

$$v_1(0) - \varepsilon p_1(0) = x_{10} + \varepsilon x_{20}, \quad p_1(1) + \varepsilon w'(v_1(1))w(v_1(1)) = 0$$

for the slow variables v_1, p_1, and two independent initial value problems

$$\varepsilon \dot{z}_2 = -z_2, \quad z_2(0) = x_{20} + p_1(0)$$

and

$$\varepsilon \dot{q}_2 = q_2, \quad q_2(1) = w(v_1(1))$$

for the fast variables z_2, q_2. The approximate expression (see (7.30)) for the control function then is

$$u = -q = -z_2 + q_2 - w(x_1) + \varepsilon[v_1 + w'(v_1)w(v_1)],$$

on noting the second of (7.44). This gives a solution of the optimal control problem (7.27)–(7.28). Here as in (7.43) $x_1 = v_1 - \varepsilon q_1 = v_1 - \varepsilon(z_2 - q_2)$, where v_1 and z_2, q_2 are the solutions of corresponding boundary value and initial value problems.

Thus, we have considered the fourth order nonlinear singularly perturbed boundary value problem which occurs in the investigation of an optimal control problem. The use of a conditionally stable slow invariant manifold in combination with specific changes of variables allows us to reduce this boundary value problem to a second order boundary value problem without singular perturbations and two independent scalar linear singularly pertsurbed initial value problems.

Chapter 8
Canards and Black Swans

Abstract The chapter is devoted to the investigation of the relationship between slow integral manifolds of singularly perturbed differential equations and critical phenomena in chemical kinetics. We consider different problems e.g., laser models, classical combustion models and gas combustion in a dust-laden medium models, 3-D autocatalator model, using the techniques of canards and black swans. The existence of canard cascades is stated for the van der Pol model and models of the Lotka-Volterra type. The language of singular perturbations seems to apply to all critical phenomena even in the most disparate chemical systems.

8.1 Introduction

A canard trajectory is a trajectory of a singularly perturbed system [129] of differential equations if it follows at first a stable invariant manifold, and then an unstable one. In both cases the length of the trajectory is more than infinitesimally small. If a trajectory at first follows an unstable invariant manifold and then a stable one, it is called a false canard. The term "canard" (or duck–trajectory) was originally introduced by French mathematicians [7, 8, 35].

In a majority of the papers devoted to canards the term "canard" is associated with periodic trajectories. However, in our work *a canard* is a one-dimensional slow invariant manifold of a singularly perturbed system of differential equations if it contains a stable (attractive) slow invariant manifold and an unstable one. It should be noted that a canard may be a result of gluing stable (attractive) and unstable (repulsive) slow invariant manifolds at one point of the breakdown surface (a subset of the slow surface which separates its stable and unstable parts) due to the availability of an additional scalar parameter in the differential system. This approach was first proposed in [59,60] and was then applied in [56,61,156,183,184].

If we have several additional parameters then we have the possibility of gluing stable (attractive) and unstable (repulsive) slow invariant manifolds at several points of the breakdown surface and thus obtain a canard cascade. If we take an additional function of a vector variable parameterizing the breakdown surface, we can glue the stable (attractive) and unstable (repulsive) slow integral manifolds at all points of the breakdown surface at the same time. As a result we obtain the continuous stable/unstable (attractive/repulsive) integral surface or *black swan*. Such surfaces

© Springer International Publishing Switzerland 2014   141
E. Shchepakina et al., *Singular Perturbations*, Lecture Notes in Mathematics 2114,
DOI 10.1007/978-3-319-09570-7_8

are considered as a multidimensional analogue of the notion of a canard. It is also possible to consider the gluing function as a special kind of partial feedback control. This can guarantee the safety of chemical regimes, even with perturbations, during a chemical process.

8.2   Singular Perturbations and Canards

Let us consider the following two-dimensional autonomous system:

$$\dot{x} = f(x, y, \mu), \tag{8.1}$$

$$\varepsilon \dot{y} = g(x, y, \mu), \tag{8.2}$$

where x, y are scalar functions of time, ε is a small positive parameter, μ is an additional scalar parameter, and f and g are sufficiently smooth scalar functions. The set of points

$$S = \{(x, y) : g(x, y, \mu) = 0\}$$

of the phase plane is called a *slow curve* of the system (8.1), (8.2).

We will need the following assumptions (1)–(3):

(1) The curve S consists of ordinary points, i.e. at every such point $(x, y) \in S$

$$[g_x(x, y, \mu)]^2 + [g_y(x, y, \mu)]^2 > 0.$$

This guarantees that S does not intersect itself.

(2) Nonregular points, i.e. points at which $g_y(x, y, \mu) = 0$, are isolated on S (all other points of S are called regular).

(3) At nonregular points, $g_{yy} \neq 0$.

Definition 5 (Jump Point). A nonregular point A (where $g_y = 0$) of the slow curve S is called a *jump point* [111] if

$$g_{yy}(A)g_x(A)f(A) > 0.$$

Definition 6 (Stable and Unstable Parts of S). The part of S which contains only regular points is called *regular*. A regular part of S, all points of which satisfy the inequality

$$g_y(x, y, \mu) < 0 \qquad \left(g_y(x, y, \mu) > 0 \right), \tag{8.3}$$

is called *stable (unstable)*.

In order be able to consider more complicated situations when a nonregular point is not ordinary (self-intersections of the slow curve, see, for instance, Example 15) or when it is ordinary but $g_{yy}(x, y, \mu) = 0$ at this point, we will use the following definition.

Definition 7 (Turning Point). A nonregular point A of the slow curve S is called a *turning point* if it separates stable and unstable parts of the slow curve.

Note that a jump point is always a turning point, but the reverse is not always true when the inequality (8.3) is not satisfied, see Fig. 8.1 and Example 14. Figure 8.1 shows a graph of a concave downwards function $x = x(y)$, which is determined by the equation $g(x, y, \mu) = 0$, hence we have $\frac{d^2 x}{dy^2}\big|_A < 0$, since A is a maximum point where $\frac{dx}{dy} = 0$. By double differentiating the equation $g(x, y, \mu) = 0$ with respect to y we obtain

$$g_{xx}\left(\frac{dx}{dy}\right)^2 + 2g_{xy}\frac{dx}{dy} + g_x\frac{d^2 x}{dy^2} + g_{yy} = 0$$

which implies

$$\frac{d^2 x}{dy^2} = -\frac{g_{xx}\left(\dfrac{dx}{dy}\right)^2 + 2g_{xy}\dfrac{dx}{dy} + g_{yy}}{g_x}.$$

From this using

$$\frac{dx}{dy}\bigg|_A = -\frac{g_y}{g_x}\bigg|_A = 0$$

we obtain

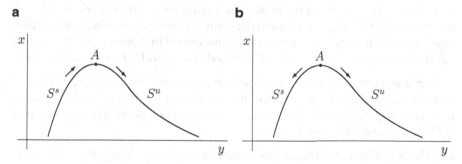

Fig. 8.1 The turning point A separates the stable (S^s) and unstable (S^u) parts of the slow curve S and (**a**) is a jump point of S; (**b**) is not a jump point because the inequality (8.3) is not satisfied. The *arrows* indicate increasing time along the slow curve

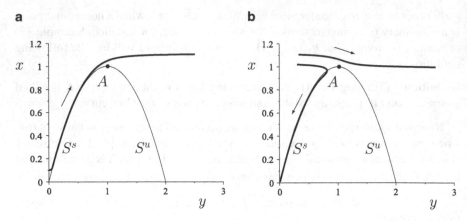

Fig. 8.2 The slow curve and the trajectories of system (8.4) with $\varepsilon = 0.01$ and (**a**) $f(x, y) = 1$; (**b**) $f(x, y) = -1$. The *arrows* indicate increasing time

$$\frac{d^2x}{dy^2}\Big|_A = -\frac{g_{yy}}{g_x}\Big|_A < 0$$

or, equivalently, $g_{yy}(A)g_x(A) > 0$.

From Fig. 8.1a it follows that the function $x = x(t)$ increases and, according to Eq. (8.1), $f(A) > 0$; the function $x = x(t)$ shown in the graph in Fig. 8.1b decreases, hence $f(A) < 0$. To illustrate this we consider

Example 14. For the system

$$\frac{dx}{dt} = f(x, y) = \pm 1, \quad \varepsilon\frac{dy}{dt} = x + y^2 - 2y \tag{8.4}$$

point $A(1, 1)$ separates the stable (S^s) and unstable (S^u) parts of the slow curve $x = 2y - y^2$ since $g_y = 2(y - 1)$, see Fig. 8.2. For $f(x, y) = 1 > 0$ we have $g_{yy}(A)g_x(A)f(A) > 0$ and the point A is a jump point. For $f(x, y) = -1 < 0$ the inequality (8.3) does not hold and the point A is not a jump point. Both of these cases appear in thermal explosion models in the case of first-order reactions: the first case takes place for $t \to -\infty$ while the second case holds for $t \to \infty$, see Sect. 7.2.

Stable and unstable parts of the slow curve are zeroth order approximations of corresponding stable and unstable slow invariant manifolds. The invariant manifolds lie in an ε-neighborhood of the slow curve, except near jump or turning points (see [117, p. 155] and references therein).

Definition 8 (Canard). Trajectories which at first move along the stable slow invariant manifold and then continue for a while along the unstable slow invariant manifold are called *canards or duck-trajectories*.

Definition 9 (False Canard). Trajectories which at first move along the unstable slow invariant manifold and then continue for a while along the stable slow invariant manifold are called *false canard trajectories*.

There is a class of problems where assumptions (1)–(3) are not fulfilled for some value of μ. Consider

Example 15. The system

$$\frac{dx}{dt} = 1, \quad \varepsilon\frac{dy}{dt} = y^2 - x^2 + \mu,$$

with $\mu = \varepsilon$ ($\mu = -\varepsilon$) has the exact slow invariant manifold $y = x$ ($y = -x$). Since, $g_y = 2y$, the part of $y = x$ ($y = -x$) with $x < 0$ ($x > 0$) is stable (attractive) while with $x > 0$ ($x < 0$) is unstable (repulsive). Note that the canard is only $y = x$. In this example [60], the point $x = 0$, $y = 0$, at which $g_x^2 + g_y^2 = 0$, is the point of self-intersection of the slow curve $y^2 - x^2 + \mu = 0$ at $\mu = 0$. Being a turning point it is not a jump point, see Definition 5.

Problems with a similar context were examined in [2, 35, 59, 60, 110]. On the one hand, this example demonstrates that canards and false canards may exist when the assumptions (1)–(3) are not fulfilled, and on the other hand, the same situation appears in thermal explosion models in the case of autocatalytic reactions. In this case the canards are the natural mathematical objects which allow us to model critical phenomena and discover critical parameter values in the form of asymptotic expansions involving powers of the small parameter ε.

8.2.1   Examples of Canards

Example 16 (Simplest Canard). As the simplest system with a canard we propose

$$\dot{x} = 1, \quad \varepsilon\dot{y} = xy + \mu.$$

Here $g(x, y) = xy + \mu$. It is clear that for $\mu = 0$, the trajectory $y = 0$ is a canard. The left part ($x < 0$) is attractive since $\frac{\partial g}{\partial y} < 0$ and the right part ($x > 0$) is repulsive. These two parts are divided by a turning point, which separates stable and unstable parts of the slow curve, at $x = 0$.

Example 17 (Simplest False Canard).
The simplest system with a false canard may be obtained by a slight modification of the previous example.

$$\dot{x} = 1, \quad \varepsilon\dot{y} = -xy + \mu.$$

Here $y = 0$ is the slow invariant manifold for $\mu = 0$. Then $g_y = -x$ and this implies that the part $x < 0$ is repulsive (or unstable) and the part $x > 0$ is attractive (or stable). Thus for $\mu = 0$, the trajectory $y = 0$ plays the role of a false canard.

The van der Pol oscillator is the most popular model used to illustrate canard trajectories. A detailed analysis can be found in [42]. We sketch the main points in the following example.

Example 18 (The van der Pol Oscillator).
    Suppose the van der Pol oscillator is biased by a constant force μ:

$$\frac{d^2y}{dt^2} + \lambda(y^2 - 1)\frac{dy}{dt} + y = \mu,$$

where μ is some real parameter, and $\lambda > 0$ as usual [194]. We write our system in Liènard form, i.e. we define $v(y)$ so that

$$\frac{d^2y}{dt^2} + \lambda(y^2 - 1)\frac{dy}{dt} = \frac{d}{dt}(\frac{dy}{dt} - \lambda v(y)),$$

which implies $v(y) = -y^3/3 + y$. We set $\lambda\frac{dx}{dt} = y - \mu$ and $\varepsilon = 1/\lambda^2$. Thus, the system becomes

$$\dot{x} = y - \mu, \quad \varepsilon\dot{y} = v(y) - x, \tag{8.5}$$

where dot denotes differentiation with respect to the new independent variable t_1, where $t_1 = t/\lambda$, i.e., $\dot{x} = \frac{dx}{dt_1} = \lambda\frac{dx}{dt}, \dot{y} = \frac{dy}{dt_1}$. The jump points $A_1(-1, -2/3)$ and $A_2(1, 2/3)$ divide the slow curve $x = v(y)$ into stable (S_1^s and S_3^s) and unstable (S_2^u) parts, see Fig. 8.3. The system (8.5) has an equilibrium at $y = \mu, x = v(y)$. Elementary analysis (see, for example, [42]) shows that the equilibrium is unstable when $-1 < \mu < 1$ and stable when $\mu > 1$ or $\mu < -1$. Indeed, the eigenvalues of the Jacobian matrix

$$J = \begin{pmatrix} 0 & 1 \\ -\varepsilon^{-1} & \varepsilon^{-1}(1 - y^2) \end{pmatrix}\Bigg|_{y=\mu}$$

are the roots of the equation

$$|J - \lambda I| = \lambda^2 - \lambda\varepsilon^{-1}(1 - \mu^2) + \varepsilon^{-1} = 0,$$

where I is an identity matrix. Hence,

$$\lambda_{1,2} = \frac{(1 - \mu^2) \pm \sqrt{(1 - \mu^2)^2 - 4\varepsilon}}{2\varepsilon}$$

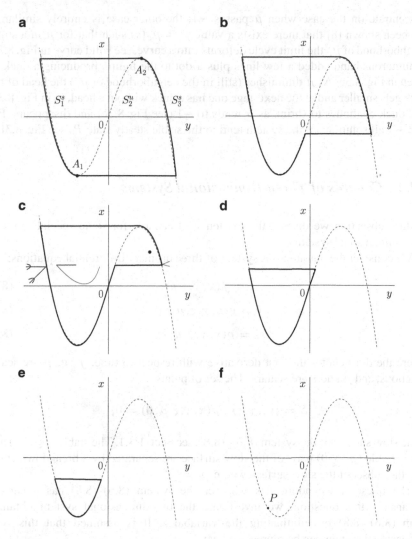

Fig. 8.3 The evolution of the limit cycle in the van der Pol oscillator for changing μ: the slow curve S (the *dashed line*) is $x = v(y)$, and the trajectories of system (8.5) are the *solid lines*. The *solid lines* constitute a limit cycle. For $\mu < -1$ the limit cycle has become a stable steady state P

are real negative numbers when $|\mu| > 1$ and $1 - \mu^2 = O(1)$ as $\varepsilon \to 0$, and $\lambda_{1,2}$ are complex numbers with negative real parts when $|\mu| > 1$ and $1 - \mu^2 = O(\varepsilon)$. The eigenvalues are real positive numbers when $|\mu| < 1$ and $1 - \mu^2 = O(1)$ as $\varepsilon \to 0$, and $\lambda_{1,2}$ are complex numbers with positive real parts when $|\mu| < 1$ and $1 - \mu^2 = O(\varepsilon)$. When $\mu \in (-1, 1)$ there will be a limit cycle, see Fig. 8.3a. When $\mu > 1$ or $\mu < -1$ there will be no limit cycle. The question is how does the limit cycle disappear when μ passes through the value -1 or 1. In what follows we shall concentrate on the case when μ passes -1; the other case is entirely similar. It has been shown [8] that there exists a value $\mu = \mu_c(\varepsilon)$ such that for μ in a small neighborhood of μ_c the limit cycle deforms into a curve, see solid curve in Fig. 8.3b. A humorous hand added a few lines plus a dot to the figure, producing a duck as given in Fig. 8.3c. As μ diminishes (still in the neighborhood of μ_c) the head of the duck gets smaller and at the next stage one has a duck without a head, as in Fig. 8.3d. The duck continues to shrink as μ tends to -1 (see Fig. 8.3e) and disappears. For $\mu < -1$ all solutions of the system tend to the stable steady state P, see Fig. 8.3f.

8.2.2   Canards of Three-Dimensional Systems

In this subsection we discuss the existence of canards for some special types of three-dimensional systems.

We consider the autonomous system of three ordinary differential equations:

$$\dot{x} = f(x, y, z, \varepsilon), \tag{8.6}$$

$$\dot{y} = g(x, y, z, \mu, \varepsilon), \tag{8.7}$$

$$\varepsilon \dot{z} = p(x, y, z, \mu, \varepsilon), \tag{8.8}$$

where the dot denotes the first derivative with respect to time, f, g, p are scalar functions, and μ and ε are scalars. The set of points

$$S = \{(x, y, z) : p(x, y, z, \mu, 0) = 0\}$$

is the slow surface of the system (8.6)–(8.8), see Sect. 1.3.1. The stable ($p_z < 0$) and the unstable ($p_z > 0$) parts of the slow surface are separated by a breakdown curve (i.e. the subset of the slow surface where $p_z = 0$).

The question we address is whether the system (8.6)–(8.8) has a canard. To answer this question, we investigate the two-dimensional system obtained from (8.6)–(8.8) by eliminating the variable t. It is assumed that this two-dimensional system can be represented as:

$$y\prime = g/f = Y(x, y, z, \mu, \varepsilon), \tag{8.9}$$

$$\varepsilon z\prime = p/f = 2xz + Z(x, y, z, \mu, \varepsilon), \tag{8.10}$$

where μ is a scalar and the function $Z(x, y, z, \mu, \varepsilon)$ has the following form:

$$Z(x, y, z, \mu, \varepsilon) = Z_1(x, y, z) + \varepsilon(C + \mu C_0) + \varepsilon Z_2(x, y, z, \mu, \varepsilon), \tag{8.11}$$

and prime represents a derivative with respect to x. Here C, C_0 are constants, functions $Y(x, y, z, \mu, \varepsilon)$, $Z_1(x, y, z)$ and $Z_2(x, y, z, \mu, \varepsilon)$ are defined, bounded and continuous in

$$\Omega = \{x \in \mathbb{R}, \ y \in \mathbb{R}, \ t|\mu + CC_0^{-1}| \le \upsilon, \ \varepsilon \in [0, \varepsilon_0]\}, \ \upsilon > 0, \varepsilon_0 > 0,$$

and satisfy the following conditions in Ω:

$$|Y(x, y, z, \mu, \varepsilon) - Y(x, \bar{y}, \bar{z}, \bar{\mu}, \varepsilon)| \le M \left(|y - \bar{y}| + |z - \bar{z}|\right) + \nu|\mu - \bar{\mu}|,$$
$$(8.12)$$

$$|Z_1(x, y, z)| \le M|z|^2, \tag{8.13}$$

$$|Z_1(x, y, z) - Z_1(x, \bar{y}, \bar{z})| \le M \left(|z| + |\bar{z}|\right)^2 |y - \bar{y}| + \frac{M}{2} \left(|z| + |\bar{z}|\right) |z - \bar{z}|,$$
$$(8.14)$$

$$|Z_2(x, y, z, \mu, \varepsilon)| \le M\nu, \tag{8.15}$$

$$|Z_2(x, y, z, \mu, \varepsilon) - Z_2(x, \bar{y}, \bar{z}, \bar{\mu}, \varepsilon)| \le M \left(|y - \bar{y}| + |z - \bar{z}|\right) + \nu|\mu - \bar{\mu}|,$$
$$(8.16)$$

where M is a positive constant and ν is a sufficiently small positive constant. These conditions are required for the existence of a unique canard which is bounded for all x and y.

The slow surface of system (8.9), (8.10) is defined by the equation $z = 0$, due to (8.11), (8.13) and the identity

$$\{2xz + Z(x, y, z, \mu, 0)\}_{z=0} \equiv 0.$$

We know (see, for instance, Sects. 2.1 and 7.2) that in an ε-neighborhood of stable and unstable foliations of the slow surface there are stable and unstable slow integral manifolds

$$z = h(x, y, \mu, \varepsilon).$$

The parameter μ ensures the existence of a gluing point of these integral manifolds, see for instance Example 19 at the end of this section. By fixing the jump point $(0, y^*)$, we can single out the trajectory

$$y = \phi(x, \mu) \quad (\phi(0, \mu) = y^*)$$

on the integral manifold $h(x, y, \mu, \varepsilon)$ which passes along the stable leaf to the jump point and then continues for a while along the unstable leaf. For convenience we use the same term 'trajectory' for both systems (8.6)–(8.8) and (8.9), (8.10). The following theorem holds.

If some natural conditions for M, ν and υ hold, then there is ε_0 such that for every $\varepsilon \in (0, \varepsilon_0)$ there exist $\mu = \mu^*(\varepsilon)$ and a canard corresponding to this parameter value which passes through the point $(0, y^*)$. The reader is referred to [184] for an exact statement and its proof.

8.2.2.1   Asymptotic Expansions for Canards

In this subsection the asymptotic expansions for the canards of the system (8.9), (8.10) are obtained.

It is assumed that functions Y and Z in (8.9), (8.10) have sufficient continuous and bounded partial derivatives with respect to all variables. For simplicity we exclude the ε-dependence of the functions Y and Z_2. Then the canard and the parameter value μ^* (corresponding to this trajectory) allow asymptotic expansions in powers of the small parameter ε:

$$\mu^* = \sum_{i \geq 0} \varepsilon^i \mu_i,$$

$$y = \phi(x, \mu^*) = \sum_{i \geq 0} \varepsilon^i \phi_i(x), \qquad\qquad (8.17)$$

$$z = \psi(x, \mu^*, \varepsilon) = h\left(x, \sum_{i \geq 0} \varepsilon^i \phi_i(x), \sum_{i \geq 0} \varepsilon^i \mu_i, \varepsilon\right) = \sum_{i \geq 0} \varepsilon^i \psi_i(x).$$

We can calculate these asymptotic expansions from (8.9), (8.10).

Note that this statements can be generalized to the cases $y \in \mathbb{R}^n, z \in \mathbb{R}$ and $y \in \mathbb{R}^n, z \in \mathbb{R}^m$.

Example 19.   As a very simple example of (8.6)–(8.8) consider the system

$$\dot{x} = 1, \quad \dot{y} = 0, \quad \varepsilon \dot{z} = 2xz + \mu - y.$$

Since $p_z = 2x$, the slow surface $2xz + \mu - y = 0$ is divided by the breakdown curve $x = 0$ into the stable part ($x < 0$) and the unstable one ($x > 0$). If μ is a parameter then the different canards are determined by

$$\dot{x} = 1, \quad y = y_0, \quad z = 0,$$

that is, they pass through the unique gluing point $x = 0, y = y_0, z = 0$ on the breakdown curve for $\mu = y_0$.

8.3   Canard Cascades

The goal of this section is to discuss the notion of *a canard cascade* as a natural generalization of the term *a canard*, to derive sufficient conditions for the existence of canard cascades, and to demonstrate how canard cascades arise in the van der Pol equation and in the singularly perturbed Lotka–Volterra model.

If it is necessary to glue stable and unstable slow invariant manifolds at several turning points, we need several additional parameters and as a result we obtain a cascade of canards or *canard cascade* [180].

In the case of a planar system, if we take an additional function whose arguments are a vector parameter and a slow variable, we can glue the stable (attractive) and unstable (repulsive) slow invariant manifolds at all breakdown points at the same time. As a result we obtain a canard cascade. It is possible to consider the gluing function as a special kind of partial feedback control. The case of *a soft control* law is studied later in Sect. 8.3.2. This implies that the control function depends on the slow variable only and it cannot change *a slow curve*.

The existence of canard cascades is studied as a problem of the gluing of stable and unstable one-dimensional slow invariant manifolds at turning points. This way of looking at the problem makes it feasible to establish the existence of canard cascades that can be considered as a generalization of canards. A further development of this approach, with applications to the van der Pol equation and a problem of population dynamics, is given later in this section.

Definition 10. The continuous slow invariant manifold of (8.1), (8.2) which contains at least two canards or false canards is called *a canard cascade*.

8.3.1 Simplest Canard Cascades

We now consider several examples of canard cascades. The differential system

$$\dot{x} = 1, \quad \varepsilon\dot{y} = x(x-1)y$$

gives a simplest canard cascade $y = 0$ which consists of two repulsive parts ($x < 0$ and $x > 1$ since $g_y = x(x-1) > 0$) and one attractive part ($0 < x < 1$ since $g_y = x(x-1) < 0$) with two turning points (breakdown points) $x = 0$ and $x = 1$.

For the next system

$$\dot{x} = 1, \quad \varepsilon\dot{y} = (x+1)x(x-1)y$$

the canard cascade $y = 0$ consists of two repulsive parts ($-1 < x < 0$ and $x > 1$) and two attractive parts ($x < -1$ and $0 < x < 1$) with three turning points $x = -1$, $x = 0$, and $x = 1$.

For a case of k turning points the following example is an obvious generalization of the two previous examples.

The system

$$\dot{x} = 1, \quad \varepsilon\dot{y} = (x - a_1)(x - a_2)\ldots(x - a_k)y$$

possesses the simplest canard cascade $y = 0$ consisting of several repulsive and attractive parts with k turning points $x_j = a_j, j = 1, \ldots, k$.

An example of a system with an infinite number of turning points is

$$\dot{x} = 1, \quad \varepsilon \dot{y} = \cos(x)y.$$

Consider now an example of a periodic canard cascade. For the planar differential system Example 9 in the Sect. 4.1, the circle $(x + \varepsilon/2)^2 + y^2 = a - \varepsilon^2/4$ is a canard. The upper semicircle is repulsive and the lower one is attractive. This canard (false canard) exists for any $a > \varepsilon^2/4$. The circle is a canard (false canard) if the movement is from the lower (upper) semicircle to the upper (lower).

8.3.2   Canard Cascade for the van der Pol Model

We consider the following generalization of the van der Pol system in the form (8.5):

$$\dot{x} = y - \mu,$$
$$\varepsilon \dot{y} = p_n(y) - x, \tag{8.18}$$

where p_n is an nth-degree polynomial in y. The corresponding slow curve $x = p_n(y)$ can have k ($k \leq n - 1$) jump points at which $p_n'(y) = 0$. If our goal is to obtain a canard cascade which contains all these points, we need to have k independent parameters. We can consider μ as a control function $\mu = \mu(x, \lambda)$ depending on the slow variable x and the k-vector $\lambda = (\lambda_1, \lambda_2, \ldots \lambda_k)$. The vector λ is a function of ε: $\lambda = \lambda(\varepsilon)$. The consideration of a variant where the function μ is a polynomial in x, i.e., $\mu = \lambda_k x^{k-1} + \lambda_{k-1} x^{k-2} + \cdots + \lambda_1$ seems quite natural. Of course, we can use a trigonometric polynomial or the linear combination of any linearly independent functions in x with the coefficients $\lambda_1, \lambda_2, \ldots \lambda_k$.

To illustrate the idea of the canard cascade construction we consider the case $n = 3$. Let $p_n(y) = Ay^3 + By^2 + Cy + D$ then $p_n'(y) = 3Ay^2 + 2By + C$. In the case $B^2 - 3AC > 0$, $p_n'(y)$ has two roots $\left(-B \pm \sqrt{B^2 - 3AC}\right)/3A$ which correspond to two jump points. It now seems natural to seek μ in the form of polynomial with $k = 2$ coefficients, i.e. $\mu = ax + b$. We search for a slow invariant manifold in the form of a polynomial $x = p_n(y) + \varepsilon q_m(y, \varepsilon)$ where $q_m(y, \varepsilon)$ is an mth-degree polynomial in y. This representation implies the equation

$$\dot{y} = -q_m(y, \varepsilon),$$

and the slow equation $\dot{x} = y - \mu$ or

$$\frac{\partial}{\partial y}\left(p_n(y) + \varepsilon q_m(y, \varepsilon)\right)\left(-q_m(y, \varepsilon)\right) = y - \mu$$

takes the form

$$\left(3Ay^2 + 2By + C + \varepsilon\frac{\partial}{\partial y}q_m(y,\varepsilon)\right)(-q_m(y,\varepsilon)) = y - ax - b, \qquad (8.19)$$

where $x = Ay^3 + By^2 + Cy + D + \varepsilon q_m(y,\varepsilon)$. Balancing the degrees of polynomials on both sides of equality (8.19) implies that $m = 1$ since the highest power of y in x is y^3, and therefore, $q_m(y,\varepsilon) = \alpha y + \beta$. Thus, we obtain the invariance equation

$$\left(3Ay^2 + 2By + C + \varepsilon\alpha\right)(-\alpha y - \beta)$$
$$= y - a\left(Ay^3 + By^2 + Cy + D + \varepsilon(\alpha y + \beta)\right) - b.$$

Equating powers of y, we obtain

$$-3\alpha A = -aA,$$

$$-3\beta A - 2\alpha B = -aB,$$

$$-2\beta B - \alpha C - \varepsilon\alpha^2 = 1 - aC - \varepsilon a\alpha,$$

$$-\varepsilon\alpha\beta - \beta C = -aD - \varepsilon a\beta - b.$$

Solving for a, b, β we get the following expressions

$$a = 3\alpha, \quad b = -3\alpha D - \varepsilon 2\alpha\beta + \beta C, \quad \beta = \frac{B}{3A}\alpha,$$

where α is a root of the quadratic equation

$$2\varepsilon\alpha^2 - 2\left(\frac{B^2}{3A} - C\right)\alpha - 1 = 0. \qquad (8.20)$$

As a result we obtain the following representation for the canard cascade

$$x = Ay^3 + By^2 + Cy + D + \varepsilon q_m(y,\varepsilon) = Ay^3 + By^2 + Cy + D + \varepsilon(\alpha y + \beta). \qquad (8.21)$$

Note that the condition $B^2 - 3AC > 0$ guarantees the existence of two jump points, with

$$\alpha = \left(\frac{B^2 - 3AC}{3A}\right)\frac{1 - \sqrt{1 + 2\varepsilon(\frac{3A}{B^2-3AC})^2}}{2\varepsilon}.$$

We do not consider the other root of Eq. (8.20)

$$\alpha = \left(\frac{B^2 - 3AC}{3A}\right) \frac{1 + \sqrt{1 + 2\varepsilon(\frac{3A}{B^2 - 3AC})^2}}{2\varepsilon}$$

because it is of order $O(1/\varepsilon)$.

In the particular case of the van der Pol system in the form (8.5) we have

$$\frac{dx}{dt} = y - \mu,$$

$$\varepsilon \frac{dy}{dt} = y - y^3/3 - x.$$

Then $A = -1/3$, $B = 0$, $C = 1$, $D = 0$ in $p_n(y)$, and, therefore, $\beta = b = 0$. Thus, we obtain

$$\mu = \mu(x, \varepsilon) = ax = 3\alpha(\varepsilon)x,$$

and, due to (8.21), the canard cascade is

$$x = Ay^3 + By^2 + Cy + D + \varepsilon(\alpha y + \beta) = y - y^3/3 + \varepsilon\alpha(\varepsilon)y,$$

where

$$\alpha(\varepsilon) = (\sqrt{1 + 2\varepsilon} - 1)/2\varepsilon.$$

Figure 8.4 shows that the canard cascade (the one-dimensional slow invariant manifold) passes near the slow curve of the system. We glue the stable and the unstable slow invariant manifolds at two jump points A_1 and A_2 (which are the extrema of the slow curve) simultaneously. Figure 8.5 demonstrates that the canard cascade plays the role of watershed line, i.e. it separates trajectories with qualitatively different behaviours.

Thus we have the following statements:

Theorem 2. *Let* $n = 3$ *and* $B^2 - 3AC > 0$ *then the differential system (8.18) possesses a canard cascade.*

Corollary 1. *The van der Pol system (8.5) possesses a canard cascade.*

In this case we have used only one additional parameter to obtain a canard cascade. Moreover, both the "canard cascade value" of this parameter $\mu = 3\alpha(\varepsilon)x$ and the canard cascade are given by exact analytical expressions.

In the general case, when the slow curve of system (8.1), (8.2) has k jump (turning) points, it is necessary to use k additional parameters to construct a canard cascade.

Fig. 8.4 Slow curve (*thin dashed line*) and canard cascade (*thick solid line*) in the case of the van der Pol equation. A_1 and A_2 are jump points where stable and unstable slow invariant manifolds are glued

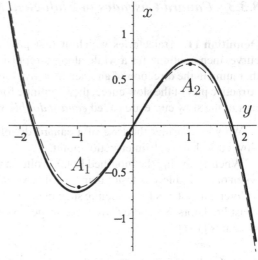

Fig. 8.5 Slow curve (*thin dashed line*), canard cascade (*thick solid line*), and two trajectories (*thin solid lines*) with different initial points I_1 and I_2 in the case of the van der Pol equation. The *arrows* indicate increasing time. The trajectory I_1 evolves to the vicinity of the upper jump point and I_2 to the vicinity of the lower jump point

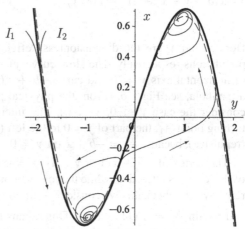

As was mentioned above, we consider control functions depending on the slow variable only. But the special case of the control function depending on the fast variable is of interest also. Eric Benoit is the author of the following statement.

Remark 8.1. The system

$$\dot{x} = \mu(y), \quad \varepsilon\dot{y} = p(y) - x$$

has a canard cascade with an invariant manifold $x = p(y) + \varepsilon q(y)$ if we choose the polynomial control $\mu(y) = -q(y)(p'(y) + \varepsilon q'(y))$. The direction of slow dynamics can be controlled by the sign of $q(y)$ to choose the sequence of true and false canards.

8.3.3 Canard Cascades in Biological Models

Definition 11. Trajectories which at first pass along an attractive part of a slow curve, then continue for a while along a repulsive part of the slow curve and after that jump in the direction of another attractive part of the slow curve, pass along this attractive part of the slow curve, then continue for a while along an another repulsive part of the slow curve are called *canard doublets*.

We will consider the case of a canard doublet [141] in the situation when the slow curve has a self-intersection point.

When analyzing the modified Lotka-Volterra model we consider the special case when one variable y is faster than the other variable x, and use singularly perturbed differential systems for modeling such phenomena. The biologically relevant case is "Fast Predators–Slow Prey" or "Fast phages–Slow bacteria", and the corresponding system is [141]:

$$\dot{x}=x(a - y - y^2 + \varepsilon\mu)=f(x,y,\varepsilon), \qquad \varepsilon\dot{y}=y(-b + x(1 + y - \delta y^2))=g(x,y).$$
$$(8.22)$$

Here x and y are the dimensionless "effective size" of the bacteria and phage populations respectively. The slow curve $y(-b + x(1 + y - \delta y^2)) = 0$ consists of horizontal axis $y = 0$ and curve $-b + x(1 + y - \delta y^2) = 0$ which looks like a parabola, see Fig. 8.6. (From the physical point of view it is not meaningful to consider the case $y < 0$, $x < 0$). The intersection point of these two lines is a turning point A_1, the part of $y = 0$ to the left (to the right) of this point is attractive (repulsive), because $\frac{\partial g}{\partial y} = -b + x$ on $y = 0$.

The vertex of $-b + x(1 + y - \delta y^2) = 0$ is a jump point A_2, the upper branch of this curve is attractive, while the lower branch until the intersection point at A_1 is repulsive. To check this we consider $\frac{\partial g}{\partial y}$ on the slow curve $-b + x(1 + y - \delta y^2) = 0$ and obtain $\frac{\partial g}{\partial y} = yx(1 - 2\delta y)$. This means that $y = 0$ is a canard and we need one additional parameter only to construct the canard cascade. Thus, for any fixed positive b and δ there exist an $a = a(\varepsilon)$, $a(0) = 1/2\delta + 1/4\delta^2$ and a canard doublet corresponding to this parameter value [141]. To calculate $a(0)$ we write the expression for the canard trajectory near the curve $-b + x(1 + y - \delta y^2) = 0$ in the form $x = \psi(y,\varepsilon) = \psi_0(y) + \varepsilon\psi_1(y) + O(\varepsilon^2)$, where $\psi_0(y) = b/(1 + y - \delta y^2)$ and we obtain

$$\varepsilon\dot{y} = y\left(-b + (\psi_0(y) + \varepsilon\psi_1(y) + O(\varepsilon^2))(1 + y - \delta y^2)\right)$$
$$= \varepsilon y(1 + y - \delta y^2)\psi_1(y) + O(\varepsilon^2).$$

The invariance equation

$$f = \frac{dx}{dt} = \frac{\partial\psi}{\partial y}\frac{dy}{dt} = \frac{\partial\psi}{\partial y}\frac{g}{\varepsilon},$$

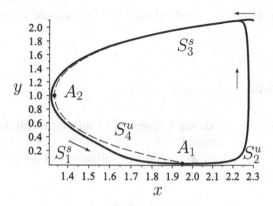

Fig. 8.6 Slow curve (*dashed line*) and the canard doublet (*solid line*) for the system (8.22). The slow curve consists of two lines: the horizontal axis $y = 0$ and the line $-b + x(1 - y - \delta y^2))$. The turning point A_1 divides the horizontal axis into the stable (attractive) part S_1^s and the unstable (repulsive) part S_2^u. The turning point A_2 divides the second line into the stable (attractive) part S_3^s and the unstable (repulsive) part S_4^u. The *arrows* indicate increasing time

i.e.

$$\frac{\partial \psi}{\partial y}\frac{g}{\varepsilon} = f,$$

with the constraint $\varepsilon = 0$, takes the form

$$\frac{\partial \psi_0}{\partial y} y(1 + y - \delta y^2)\psi_1(y) = \psi_0(y)(a(0) - y - y^2)$$

to lowest order. It is easy to obtain the "canard value" of $a(0)$ from the continuity condition for $\psi_1(y)$ at $y = 1/2\delta$:

$$\psi_1(y) = \frac{\psi_0(y)(a(0) - y - y^2)}{\frac{\partial \psi_0}{\partial y} y(1 + y - \delta y^2)} = -\frac{a(0) - y - y^2}{y(1 - 2\delta y)},$$

since we require $a(0) - y - y^2 = 0$ at $y = 1/2\delta$. The canard doublet trajectory at first passes along an attractive part of $-b + x(1 + y - \delta y^2) = 0$, then continues for a while along a repulsive part of this curve and after that jumps in the direction of attractive part of $y = 0$, passes along this attractive part, then continues for a while along a repulsive part of $y = 0$, see Fig. 8.6.

A canard doublet of the same form was discovered in a forest pest model [22]. For small values of the timescale of the young trees, the model can be reduced to a two-dimensional model. This model allows oscillations where long pest-free periods are interspersed with outbreaks of high pest concentration. In this case the variable

x corresponds to the population of old trees and the variable y corresponds to the pest population.

8.4 Black Swans

In this section we use the standard approach of integral manifolds that we have developed to study slow integral surfaces of variable stability (or black swans). These surfaces are considered as natural generalizations of the notion of a canard. We consider the system

$$\dot{x} = f(x, y, z, \varepsilon), \tag{8.23}$$

$$\dot{y} = g(x, y, z, \mu, \varepsilon), \tag{8.24}$$

$$\varepsilon\dot{z} = p(x, y, z, \mu, \varepsilon), \tag{8.25}$$

where ε is a small positive parameter, μ is a scalar parameter, x and z are scalar variables, y is a vector of dimension n.

Recall that the slow surface S of system (8.23)–(8.25) is the surface described by the equation

$$p(x, y, z, \mu, 0) = 0. \tag{8.26}$$

Let $z = \phi(x, y)$ be an isolated solution of Eq. (8.26), then the subset S^s (S^u) of S defined by

$$\frac{\partial p}{\partial z}(x, y, \phi(x, y), \alpha, 0) < 0 \ \ (> 0)$$

is the stable (unstable) subset of S. The subset of S defined by

$$\frac{\partial p}{\partial z}(x, y, \phi(x, y), \alpha, 0) = 0$$

is the breakdown surface (the breakdown curve if dim $y = 1$) with dimension equal to dim y.

In ε-neighborhood of the subset S^s, which is part of a slow surface, (S^u) there exists a stable (unstable) slow invariant manifold. If the stable and unstable slow invariant manifolds are glued at all points of the breakdown surface then the system has a continuous invariant surface which is called the black swan.

The term "black swan" is suggested for two reasons. The first is that a swan is a bird from the family of ducks. The second is connected with the usual meaning of "black swan" in the sense of a rare phenomenon. It should be noted also that the French term "canard" is used in the sense of a false rumour[1] in English.

An additional parameter is used to glue together the stable and unstable parts of a canard, and we need an additional function to glue integral manifolds whose dimension is greater than one. The argument of this function is a vector variable parameterizing the breakdown surface.

The term "black swan" can also be extended to the case where $\dim z > 1$ [155]. As an example of a black swan we return to Example 19: $\dot{x} = 1$, $\dot{y} = 0$, $\varepsilon\dot{z} = 2xz + \mu - y$. If μ is a function of the variable y then for $\mu = y$ the invariant manifold $z = 0$ is stable for $x < 0$ and unstable for $x > 0$ since $p = 2xz$. This means that the plane $z = 0$ is a black swan, because $z = 0$ is a slow invariant manifold of variable stability.

We consider the system (8.23)–(8.25) reduced to the form

$$\frac{dy}{dx} = Y(x, y, z, \varepsilon), \qquad y \in \mathbb{R}^n, \qquad x \in \mathbb{R}; \tag{8.27}$$

$$\varepsilon\frac{dz}{dx} = 2xz + \mu + Z(x, y, z, \mu, \varepsilon), \qquad |z| \le r, \tag{8.28}$$

where r is a positive constant, μ and ε are scalars. It is supposed that the functions Y, Z are continuous and satisfy some natural smoothness and boundedness conditions (see, for example [117]).

We consider μ as a function: $\mu = \mu(y, \varepsilon)$.

Continuing in the same manner as in Sect. 2.3 we can obtain the asymptotic representations for black swans and the corresponding functions $\mu(y, \varepsilon)$. We will give some examples later.

If a gluing function $\mu(y, \varepsilon)$ exists, then every trajectory on the slow integral manifold is a canard if it crosses the surface $x = 0$ from the stable part ($x < 0$) to unstable part ($x > 0$). Thus, in Example 19 for $\mu = y$, every trajectory on the slow integral manifold $z = 0$ is a canard. In the case when the gluing function has to be a constant, $\mu(y, \varepsilon) = \mu(y_0, \varepsilon)$, the stable and unstable parts of the integral manifold can be glued at one point $y = y_0$ only. The canard passes only through this point.

8.5 Laser and Chemical Models

In this section we shall consider the relationship between canards and black swans and critical phenomena in different laser and chemical systems. We shall show that canards play the role of *separating solutions*. This means that canards simulate the critical regimes separating the basic types of chemical regimes, e.g., slow from explosive regimes.

---

[1]"An absurd story circulated as a hoax", see Shorter Oxford English Dictionary.

The application of black swans consisting entirely of canards to the modelling of critical phenomena permits us to take into account small perturbations in the chemical systems. Moreover we can use black swans for the modelling of critical phenomena in problems without fixed initial conditions.

Before we consider combustion models, we first give some relatively simple examples of other physical systems.

8.5.1   Lang–Kobayashi Equations

External cavity semiconductor lasers present many interesting features for both technological applications and fundamental non-linear science. Their dynamics have been the subject of numerous studies for the last twenty years. Motivations for these studies vary from the need for stable tunable laser sources, for laser cooling or multiplexing, to the general understanding of their complex stability and chaotic behavior. The typical experiment is usually described by a set of delay differential equations introduced by Lang and Kobayashi [97]:

$$\dot{E} = \kappa \left(1 + i\alpha\right) \left(N - 1\right) E + \gamma e^{-i\varphi_0} E\left(t - \tau\right),$$
$$\dot{N} = -\gamma_{\parallel} \left(N - J + |E|^2 N\right). \tag{8.29}$$

Here E is the complex amplitude of the electric field, N is the carrier density, J is pumping current, κ is the field decay rate, $1/\gamma_{\parallel}$ is the spontaneous time scale, α is the linewidth enhancement factor, γ represents the feedback level, φ_0 is the phase of the feedback if the laser emits at the solitary laser frequency and τ is the external cavity round trip time. There are few analytical results since delay equations are nonlocal. However, this model was recently reduced to a 3D dynamical system describing the temporal evolution of the laser power $P = |E|^2$, carrier density N and phase difference $\eta(t) = \varphi(t) - \varphi(t - \tau)$. This was achieved by assuming $P(t - \tau) = P(t)$ together with the approximation $\dot{\varphi} = \eta/\tau + \dot{\eta}/2$. This expression remains valid when the phase fluctuates on a time scale much shorter than the re-injection time τ. Under these approximations, the Lang–Kobayashi equations (8.29) reduce to [78]:

$$\dot{P} = 2\left[\kappa\left(N - 1\right) + \gamma \cos\left(\eta + \varphi_0\right)\right] P,$$
$$\dot{N} = -\gamma_{\parallel}\left(N - J + PN\right), \tag{8.30}$$
$$\dot{\eta} = 2\left[-\frac{\eta}{\tau} + \kappa\alpha\left(N - 1\right) - \gamma \sin\left(\eta + \varphi_0\right)\right].$$

This model was successfully used to describe low frequency fluctuations commonly observed in semiconductor lasers with optical feedback [78].

Suppose that the following relations hold for the various parameters in the model:

$$\gamma = o(1),\ 1/\tau = o(1), \gamma_{\|} = o(1);\ \kappa = O(1),\ \alpha = O(1).$$

In this case system (8.30) is singular singularly perturbed, see Sect. 5. With new variables and parameters:

$$\varepsilon = \tau_0/\tau,\ t_1 = \varepsilon t,\ x = \ln P - \eta/\alpha,\ \gamma = \varepsilon\gamma_0,\ \gamma_{\|} = \varepsilon\gamma_{\|0},$$

where $\varepsilon = o(1)$ is a dimensionless parameter, $\gamma_0 = O(1)$, $1/\tau_0 = O(1)$, $\gamma_{\|0} = O(1)$, the system (8.30) takes the form

$$\varepsilon \dot{P} = 2\left[\kappa\,(N - 1) + \varepsilon\gamma_0 \cos\,(\alpha \ln P - \alpha x + \varphi_0)\right] P,$$
$$\dot{N} = -\gamma_{\|0}\,(N - J + PN),\qquad\qquad\qquad\qquad (8.31)$$
$$\dot{x} = 2\gamma_0 \left[\cos\,(\alpha \ln P - \alpha x + \varphi_0) + \frac{1}{\alpha}\sin\,(\alpha \ln P - \alpha x + \varphi_0)\right] + 2\frac{\eta}{\alpha\tau_0},$$

where the dot indicates differentiation w.r.t. t_1. In this case, system (8.31) possesses an exact slow invariant manifold $P \equiv 0$, which coincides with the slow surface of the system, since P is outside the square bracket and $N \neq 1$. The exchange of stability on this surface is carried out on the breakdown curve

$$\frac{\partial p}{\partial P}(x, N, 0, 0) = 2\kappa(N - 1) = 0,$$

where $p(x, N, 0, \varepsilon) = 2\left[\kappa\,(N - 1) + \varepsilon\gamma_0 \cos\,(\alpha \ln P - \alpha x + \varphi_0)\right] P$. Hence $P \equiv 0$ is the black swan of (8.31); the part with $N < 1$ is stable while the part with $N > 1$ is unstable. Trajectories of the system (8.31) are the spirals containing a stable/unstable part of the slow motion along $P \equiv 0$, see Fig. 8.7, i.e. they are canards. This behavior corresponds to a pulsed operation of the laser, when the optical power appears in pulses of some duration at some repetition rate.

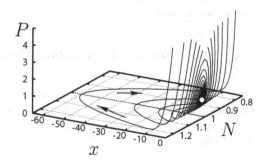

Fig. 8.7 The black swan $P \equiv 0$ and the trajectories of system (8.31). The *arrows* indicate increasing time

8.5.2   The Simple Laser

The nonlinear first-order equation

$$\dot{y} = ky^p + \lambda(t)y + \delta, \quad 0 < \varepsilon \ll 1, \quad \delta \leq 0,$$

with $k = \pm 1$, $p = 2,3$ and the control parameter $\lambda(t) = \lambda_0 + \varepsilon t$, $\lambda_0 < 0$, is a typical model of simple lasers, and lasers with saturable absorbers, where y is a dimensionless amplitude of the field [103]. The additional parameter δ characterizes the magnitude of the imperfections, ε is small quantity, and $\lambda_0 = O(1)$. Note that these equations with $\lambda(t) = \lambda_0 + \varepsilon t$ may be written in the form

$$\dot{\lambda} = \varepsilon, \quad \dot{y} = ky^p + \lambda y + \delta.$$

For $\delta = 0$ this system has the canard $y = 0$. Physically the canard simulates the critical regime separating the basic types of the regimes, slow and self-accelerating [117, Chap. 8]. We consider in more detail the case $p = 3, k = -1$:

$$\dot{\lambda} = \varepsilon, \quad \dot{y} = -y^3 + \lambda y + \delta. \tag{8.32}$$

The slow curve S is described by the equation $-y^3 + \lambda y + \delta = 0$ and has a different form depending on δ (see Fig. 8.8). When $\delta < 0$ ($\delta > 0$) the trajectories of the system move along the stable part S_1^s (S_3^s) of the slow curve, see Fig. 8.8a, b. In these cases the trajectories describe fundamentally different slow regimes, with either a monotonically increasing or decreasing amplitude.

When $\delta = 0$ system (8.32) has the exact canard $y \equiv 0$, see Fig. 8.8c. Other canards, which are the intermediate trajectories in the region between those shown above in Fig. 8.8a, b, pass along the stable part S_1^s and then along the unstable part S_2^u, jump from the slow invariant manifold towards the stable part S_3^s and then move along it, see Fig. 8.8c. This means that the amplitude of the field, having remained close to zero for a long time, almost instantaneously jumps to attain significance and then increases slowly.

8.5.3   The Classical Combustion Model

From the mathematical viewpoint the situation which appears in the model of an autocatalytic reaction looks like Example 15 in Sect. 8.2.

The system showing the autocatalytic features of the reaction is given by (7.17) and (7.18) viz.,

$$\varepsilon \frac{d\theta}{d\tau} = \eta(1 - \eta)\exp\left(\theta/\left(1 + \beta\theta\right)\right) - \alpha\theta, \tag{8.33}$$

$$\frac{d\eta}{d\tau} = \eta(1 - \eta)\exp\left(\theta/\left(1 + \beta\theta\right)\right), \tag{8.34}$$

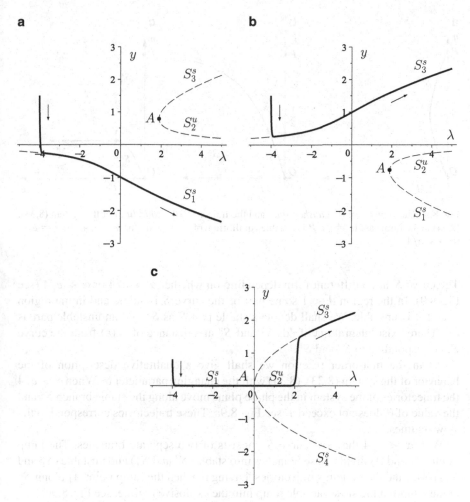

Fig. 8.8 The slow curve S (the *dashed line*) and the trajectory (the *solid line*) of system (8.32) for $\varepsilon = 0.1$, $\lambda_0 = -4$ and (**a**) $\delta = -1$, (**b**) $\delta = 1$, (**c**) $\delta = 0$. The parts S_1^s, S_3^s and S_4^s arc stable while S_2^u is unstable, A is a jump point in each case

with the initial conditions

$$\eta(0) = \eta_0/ (1 + \eta_0) = \bar{\eta}_0, \quad \theta(0) = 0.$$

To simplify the demonstration of the main qualitative effects we use a widespread assumption in thermal explosion theory, $\beta = 0$. A detailed analysis shows that the result is little different from the case $\beta \neq 0$. In this case the slow curve S of the system (8.33), (8.34) is described by the equation

$$\eta(1 - \eta)e^\theta - \alpha\theta = 0.$$

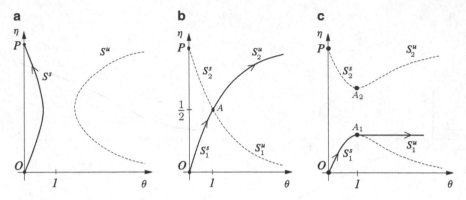

Fig. 8.9 The slow curve (the *dashed line*) and the trajectory (the *solid line*) of the system (8.33), (8.34) in the limit case ($\varepsilon = 0$). P is a stable equilibrium of the system. (**a**) $\alpha > e/4$, (**b**) $\alpha = e/4$, (**c**) $\alpha < e/4$

The curve S has a different form depending on whether $\alpha > e/4$ or $\alpha < e/4$ (see Fig. 8.9). In the region $\theta < 1$ some part of the curve S is stable and in the region $\theta > 1$ it is unstable. We shall denote a stable part S as S^s and an unstable part as S^u. There exist integral manifolds S_ε^s and S_ε^u at a distance of $O(\varepsilon)$ from the curve S, corresponding to S^s and S^u.

As in the first-order reaction we shall give a qualitative description of the behavior of the system (8.33), (8.34) with the changing parameter α. When $\alpha > e/4$ the trajectories of the system in the phase plane move along the stable branch S^s and the value of θ does not exceed 1, see Fig. 8.9a. These trajectories correspond to the slow regimes.

With $\alpha < e/4$ the slow curve S consists of two separate branches. The jump points A_1 and A_2 divide these branches into stable (S_1^s and S_2^s) and unstable (S_1^u and S_2^u) parts, and the system's trajectories, having reached the jump point A_1 along S_1^s at the tempo of the slow variable jump into the explosive regime, see Fig. 8.9c.

Due to the continuous dependence of the right-hand side of (8.33), (8.34) on the parameter α there are some intermediate trajectories in the region between those shown above in Fig. 8.9a, b in the neighborhood of $\alpha = e/4$, and also a critical trajectory. With $\alpha = e/4$ the slow curve S has a self-intersection point $A(1, 1/2)$.

The canard, passing along the stable part of slow curve S_1^s and then along the unstable part S_2^u at some value of α (see Fig. 8.9b), is taken as a mathematical object to model the critical trajectory, which corresponds to a chemical reaction separating the domain of self-acceleration reactions, i.e. the explosive reactions, ($\alpha < \alpha^*$) and the domain of non-explosive reactions ($\alpha > \alpha^*$). The critical value of the parameter $\alpha = \alpha^*$ corresponding to this trajectory is found in the form

$$\alpha^* = \alpha_0 + \varepsilon\alpha_1 + \ldots , \text{ where } \alpha_0 = e/4. \tag{8.35}$$

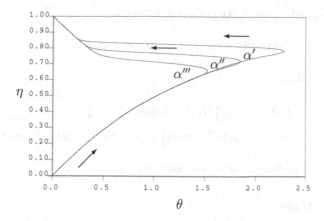

Fig. 8.10 Canard trajectories of system for $\varepsilon = 0.05$, $\alpha' = 0.659941603$, $\alpha'' = 0.659941646$, $\alpha''' = 0.659952218$

Note that there is one more trajectory passing along S_ε^u and S_ε^s in Fig. 8.9b. This trajectory, passing along S_1^u and then along S_2^s, is a false canard which does not correspond to any chemical regime. The value $\alpha = \alpha^{**}$ corresponds to this trajectory. At $\alpha > \alpha^{**}$ we get a region of slow regimes and the trajectories of system (8.33), (8.34) will pass along the stable part of slow curve, see Fig. 8.9a.

The transition trajectories between S_ε^s and S_ε^u correspond to the interval (α^*, α^{**}), see Fig. 8.10 and note $\alpha', \alpha'', \alpha''' \in (\alpha^*, \alpha^{**})$. To calculate the critical value of the parameter $\alpha = \alpha^*$ (and $\alpha = \alpha^{**}$) we substitute (8.35) and the expression for the corresponding canard [59, 60]

$$\eta = H(\theta, \varepsilon) \equiv H_0(\theta) + \varepsilon H_1(\theta) + \dots$$

into (8.33), (8.34). We write (8.33), (8.34) as

$$\varepsilon \frac{d\eta}{d\tau} = \varepsilon \frac{d\eta}{d\theta} \frac{d\theta}{d\tau} = \frac{d\eta}{d\theta}[\eta(1 - \eta)e^\theta - \alpha\theta]$$

to get

$$\bigl(H(\theta, \varepsilon)\,(1 - H(\theta, \varepsilon))\,e^\theta - \alpha(\varepsilon)\theta\bigr)\,H'(\theta, \varepsilon) = \varepsilon H(\theta, \varepsilon)\,(1 - H(\theta, \varepsilon))\,e^\theta$$

or, in more detailed form,

$$\Biggl((H_0(\theta) + \varepsilon H_1(\theta) + \dots)\,(1 - H_0(\theta) - \varepsilon H_1(\theta) - \dots)\,e^\theta$$

$$-(\alpha_0 + \varepsilon\alpha_1 + \dots)\theta\Biggr)\Bigl[H_0'(\theta) + \varepsilon H_1'(\theta) + \dots\Bigr]$$

$$= \varepsilon\,(H_0(\theta) + \varepsilon H_1(\theta) + \dots)\,(1 - H_0(\theta) - \varepsilon H_1(\theta) - \dots)\,e^\theta.$$

Equating the coefficients of like powers of ε we get

$$g(H_0(\theta), \theta) = H_0(1 - H_0)e^\theta - \alpha_0\theta = 0, \tag{8.36}$$

since $H_0'(\theta) \neq 0$, and

$$H_0(1 - H_0)e^\theta = H_0'\left[H_1(1 - 2H_0)e^\theta - \alpha_1\theta\right]$$
$$+H_1'\left[H_0(1 - H_0)e^\theta - \alpha_0\theta)\right] = H_0'\left[H_1(1 - 2H_0)e^\theta - \alpha_1\theta\right],$$

using (8.36). From these equations we obtain

$$H_0(\theta) = \frac{1}{2} \pm \sqrt{\frac{1}{4} - \alpha_0\theta e^{-\theta}},$$

$$H_1(\theta) = \frac{H_0(1 - H_0)e^\theta + \alpha_1\theta H_0'}{H_0'(1 - 2H_0)e^\theta} = \frac{\theta(\alpha_1 H_0' + \alpha_0)}{H_0'(1 - 2H_0)e^\theta}.$$

The coefficients in the expression (8.35) α_i ($i = 0, 1, \ldots$) are found by requiring the functions $H_i = H_i(\theta)$ to be continuous at $\theta = 1$. We note from the form of $H_1 = H_1(\theta)$ it has a zero under the line when $\theta = 1$. Thus we require $\alpha_1 = -\alpha_0/H_0'(1)$, where the value $H_0'(1)$ can be found from Eq. (8.36) after double differentiation with respect to θ:

$$\left\{g_{\theta\theta} + 2g_{\theta H_0} + g_{H_0 H_0}\left(H_0'\right)^2 + g_{H_0} H_0''\right\}\Big|_{\theta=1} = 0,$$

or by expanding about $\theta = 1$, to give $H_0'(1) = \pm\sqrt{\alpha_0/2e}$. Hence

$$H_0'(1) = \pm\frac{1}{2\sqrt{2}}.$$

We take "$+$" for the canard, because the function $H_0(\theta)$ monotonically increases, and "$-$" for the false canard, because $H_0(\theta)$ monotonically decreases in this case. Thus, we have

$$\alpha^* = e/4(1 - 2\sqrt{2}\varepsilon) + O(\varepsilon^2),$$

$$\alpha^{**} = e/4(1 + 2\sqrt{2}\varepsilon) + O(\varepsilon^2).$$

Note that critical regimes of combustion were investigated in [1, 3, 37, 43, 52, 55, 57, 89, 107, 108, 116, 165, 166, 193]

8.5.4 Canards and Black Swan in a Model of a 3-D Autocatalator

In this section a two-dimensional stable–unstable slow integral manifold (black swan), consisting entirely of canards, which simulates the critical phenomena for different initial data of the dynamical system, is constructed. It is shown that this procedure leads to the phenomenon of auto-oscillations in the chemical system. The application of a black swan permits us to take into account small perturbations in the chemical systems.

A model of a three-dimensional autocatalator has the form [135, 137]:

$$\frac{dx}{d\tau} = \mu(5/2 + y) - xz^2 - x,$$

$$\frac{dy}{d\tau} = z - y, \qquad\qquad\qquad (8.37)$$

$$\varepsilon\frac{dz}{d\tau} = xz^2 + x - z,$$

where

$$x \geq 0, \quad y \geq 0, \quad z \geq 0, \quad 0 \leq \mu < 1. \qquad\qquad (8.38)$$

The system (8.37) simulates a sort of Belousov–Zhabotinsky reaction. The variables x, y and z represent dimensionless concentrations of three chemical reagents, ε is a small positive parameter, μ is a bifurcation parameter.

The slow surface (see Fig. 8.11) of the system (8.37) is described by the equation

$$F(x, y, z) = xz^2 + x - z = 0.$$

The breakdown surface, which is described by

$$F = 0, \quad \frac{\partial F}{\partial z} = 2xz - 1 = 0,$$

consists of two straight lines, but only one

$$x = 0.5, \quad z = 1 \qquad\qquad\qquad (8.39)$$

has physical meaning. The other is $x = -0.5$, $z = -1$ and these violate (8.38). The breakdown surface divides the slow surface into three leaves S_1^u ($z > 1$), S_2^u ($z < 1$), S^s ($|z| < 1$, see Fig. 8.11), which are zeroth order approximations for the corresponding slow integral manifolds $S_{1,\varepsilon}^u$, $S_{2,\varepsilon}^u$ and S_ε^s. Manifolds $S_{1,\varepsilon}^u$ and $S_{2,\varepsilon}^u$ are unstable and S_ε^s is stable. Note that the part of S_ε^s with $0 \leq x < 0.5$ and $S_{1,\varepsilon}^u$ are situated in the domain of interest as given by (8.38).

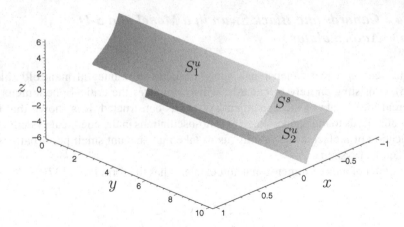

Fig. 8.11 The slow surface of the system (8.37)

The system (8.37) has an equilibrium at P, where $\frac{dx}{d\tau} = \frac{dy}{d\tau} = \frac{dz}{d\tau} = 0$, given by

$$\left(\frac{10\mu(1-\mu)}{29\mu^2 - 8\mu + 4}, \frac{5\mu}{2(1-\mu)}, \frac{5\mu}{2(1-\mu)} \right),$$

and with $\mu = 0$ this equilibrium is a stable node and lies at $(0,0,0)$. In [158] it has been shown that P is a stable equilibrium (node) on the stable leaf of the slow surface when $0 \leq \mu < 2/7$ and it is an unstable equilibrium (saddle) on the unstable leaf S_1^u when $\mu > 2/7$.

The slow surface is an approximation to a slow integral manifold (for $\varepsilon = 0$), hence it is possible to determine the basic types of chemical regimes and corresponding values of the control parameter μ.

With $0 \leq \mu < 2/7$ a trajectory of the system (8.37), starting from an initial point in the basin of attraction of the stable slow integral manifold S_ε^s, follows S_ε^s for a short time and tends to the stable equilibrium P as $\tau \to \infty$, see Fig. 8.12a. This behavior corresponds to the slow chemical regime.

With $\mu > 2/7$ a trajectory of (8.37) will follow the S_ε^s to the breakdown line (8.39). After this time, $z(\tau)$ will increase rapidly, see Fig. 8.12b. This behavior characterizes the explosive regime. The point P in Fig. 8.12b is the equilibrium point $(0.4, 1.667)$ of the system corresponding to $\mu = 0.4$. It lies on the slow surface above the breakdown curve whose xOz-projection is $(0.5, 1)$.

Due to the continuous dependence of the right-hand side of (8.37) on the parameter μ there is a critical trajectory in the neighborhood of $\mu = 2/7$, which separates the two regions of the chemical reactions described above.

The availability of the additional scalar parameter μ provides the possibility of gluing S_ε^s and $S_{1,\varepsilon}^u$ at one point of the breakdown line (8.39). The canard trajectory passes through this point.

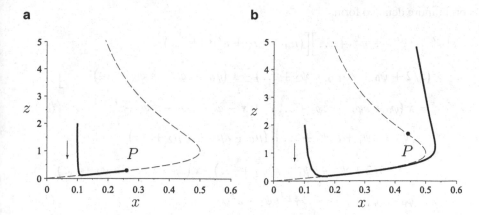

Fig. 8.12 xOz-projection of the slow surface (*dashed line*) and the trajectory (*solid line*) of system (8.37); $\varepsilon = 0.01$ and (**a**) $\mu = 0.1$, (**b**) $\mu = 0.4$. The *arrows* indicate increasing time

The canard plays the role of a separating solution, and is taken as a mathematical object to model the critical trajectory corresponding to the critical value $\mu = \mu^* = 2/7 + O(\varepsilon)$, $(\varepsilon \to 0)$. This means that the canard simulates a chemical reaction separating the domain of self-accelerating reactions $(\mu > \mu^*)$ and the domain of slow reactions $(\mu < \mu^*)$.

8.5.4.1   Canard in the 3-D Autocatalator

We can find the canard solution, and corresponding critical value of parameter $\mu = \mu^*$, by the following asymptotic expansions

$$z = z(x, \varepsilon) = \varphi_0(x) + \varepsilon\varphi_1(x) + \varepsilon^2\varphi_2(x) + \dots , \tag{8.40}$$

$$y = y(x, \varepsilon) = \psi_0(x) + \varepsilon\psi_1(x) + \varepsilon^2\psi_2(x) + \dots , \tag{8.41}$$

$$\mu^* = \mu(\varepsilon) = \mu_0 + \varepsilon\mu_1 + \varepsilon^2\mu_2 + \dots . \tag{8.42}$$

From (8.37) and (8.40)–(8.42) using the invariance equation

$$\varepsilon\frac{dz}{dx}\frac{dx}{d\tau} = \varepsilon\frac{dz}{d\tau}, \quad \frac{dy}{dx}\frac{dx}{d\tau} = \frac{dy}{d\tau},$$

we have

$$\varepsilon z'(x, \varepsilon)\Big[\mu\,(5/2 + y(x, \varepsilon)) - xz^2(x, \varepsilon) - x\Big] = xz^2(x, \varepsilon) + x - z(x, \varepsilon),$$

$$y'(x, \varepsilon)\Big[\mu\,(5/2 + y(x, \varepsilon)) - xz^2(x, \varepsilon) - x\Big] = z(x, \varepsilon) - y(x, \varepsilon)$$

or, in more detailed form,

$$\varepsilon\left[\varphi_0' + \varepsilon\varphi_1' + \varepsilon^2\varphi_2' + \ldots\right]\left[\left(\mu_0 + \varepsilon\mu_1 + \varepsilon^2\mu_2 + \ldots\right)\right.$$

$$\left.\times\left(5/2 + \psi_0 + \varepsilon\psi_1 + \varepsilon^2\psi_2 + \ldots\right) - x\left(\varphi_0 + \varepsilon\varphi_1 + \varepsilon^2\varphi_2 + \ldots\right)^2 - x\right]$$

$$= x\left(\varphi_0 + \varepsilon\varphi_1 + \varepsilon^2\varphi_2 + \ldots\right)^2 + x - \varphi_0 - \varepsilon\varphi_1 - \varepsilon^2\varphi_2 - \ldots, \qquad (8.43)$$

$$\left[\psi_0' + \varepsilon\psi_1' + \varepsilon^2\psi_2' + \ldots\right]\left[\left(\mu_0 + \varepsilon\mu_1 + \varepsilon^2\mu_2 + \ldots\right)\right.$$

$$\left.\times\left(5/2 + \psi_0 + \varepsilon\psi_1 + \varepsilon^2\psi_2 + \ldots\right) - x\left(\varphi_0 + \varepsilon\varphi_1 + \varepsilon^2\varphi_2 + \ldots\right)^2 - x\right]$$

$$= \varphi_0 + \varepsilon\varphi_1 + \varepsilon^2\varphi_2 - \psi_0 - \varepsilon\psi_1 - \varepsilon^2\psi_2 + \ldots. \qquad (8.44)$$

Setting $\varepsilon = 0$ in (8.43), (8.44) we obtain

$$x\varphi_0^2 - \varphi_0 + x = 0,$$

$$\psi_0'\left[\mu_0\left(5/2 + \psi_0\right) - x\varphi_0^2 - x\right] = \varphi_0 - \psi_0,$$

or

$$x\varphi_0^2 - \varphi_0 + x = 0, \qquad (8.45)$$

$$\psi_0'\left[\mu_0\left(5/2 + \psi_0\right) - \varphi_0\right] = \varphi_0 - \psi_0. \qquad (8.46)$$

Then Eq. (8.45) defines the function $\varphi_0 = \varphi_0(x)$, the first term in z, and Eq. (8.46) defines the function $\psi_0 = \psi_0(x)$, the first term in y.

We now equate terms in ε^1 in (8.43), (8.44):

$$\varphi_0'\left[\mu_0\left(5/2 + \psi_0\right) - x\varphi_0^2 - x\right] = \varphi_1(2x\varphi_0 - 1),$$

$$\psi_1'\left[\mu_0\left(5/2 + \psi_0\right) - x\varphi_0^2 - x\right] + \psi_0'\left[\mu_1\left(5/2 + \psi_0\right) + \mu_0\psi_1 - 2x\varphi_0\varphi_1\right] = \varphi_1 - \psi_1$$

or, taking into account (8.45),

$$\varphi_0'\left[\mu_0\left(5/2 + \psi_0\right) - \varphi_0\right] = \varphi_1(2x\varphi_0 - 1), \qquad (8.47)$$

$$\psi_1'\left[\mu_0\left(5/2 + \psi_0\right) - \varphi_0\right] + \psi_0'\left[\mu_1\left(5/2 + \psi_0\right) + \mu_0\psi_1 - 2x\varphi_0\varphi_1\right] = \varphi_1 - \psi_1. \qquad (8.48)$$

On the breakdown surface (8.39) we have $2x\varphi_0 - 1 = 0$. By continuity of the function $\varphi_1 = \varphi_1(x)$ we thus require the following condition from (8.47)

$$\mu_0\left(5/2 + \psi_0\left(0.5\right)\right) - \varphi_0\left(0.5\right) = 0.$$

From this and (8.39) we obtain

$$\mu_0 = \frac{1}{5/2 + \psi_0(0.5)}, \tag{8.49}$$

since $\varphi_0(0.5) = 1$. Next, equating terms in ε^2 in (8.43), (8.44) and applying (8.45), we get

$$\varphi_0'\left[\mu_1(5/2 + \psi_0) + \mu_0\psi_1 - 2x\varphi_0\varphi_1\right] + \varphi_1'\left[\mu_0(5/2 + \psi_0) - \varphi_0\right]$$

$$= x\varphi_1^2 + (2x\varphi_0 - 1)\varphi_2, \tag{8.50}$$

$$\psi_0'\left[\mu_0\psi_2 + \mu_1\psi_1 + \mu_2(5/2 + \psi_0) - 2x\varphi_0\varphi_2 - x\varphi_1^2\right]$$

$$+ \psi_1'\left[\mu_1(5/2 + \psi_0) + \mu_0\psi_1 - 2x\varphi_0\varphi_1\right] + \psi_2'\left[\mu_0(5/2 + \psi_0) - \varphi_0\right] = \varphi_2 - \psi_2.$$

On the breakdown line (8.39) the coefficient of the function $\varphi_2 = \varphi_2(x)$ in (8.50) is equal to zero. To avoid a discontinuity in this function we require, taking into account (8.49), the following condition (the remaining terms sum to zero)

$$\varphi_0'(0.5)\left[\mu_1(5/2 + \psi_0(0.5)) + \mu_0\psi_1(0.5) - \varphi_0(0.5)\varphi_1(0.5)\right] = 0.5\varphi_1^2(0.5). \tag{8.51}$$

To calculate the value $\varphi_0'(0.5)$ we differentiate (8.45) with respect to x:

$$\varphi_0^2 + \varphi_0'\left(2x\varphi_0 - 1\right) + 1 = 0.$$

The coefficient at φ_0' in this expression is equal to zero on the breakdown surface. Therefore, differentiating the last equation with respect to x, we have

$$2\varphi_0'\left(2\varphi_0 + x\varphi_0'\right) + \varphi_0''\left(2x\varphi_0 - 1\right) = 0.$$

On the breakdown line $2x\varphi_0 - 1 = 0$ and then $\varphi_0' = -2\varphi_0/x$, i.e.

$$\varphi_0'(0.5) = -4,$$

on the breakdown line. Substituting this value into (8.51) we find

$$\mu_1 = \frac{\varphi_1(0.5)[1 - \varphi_1(0.5)/8] - \mu_0\psi_1(0.5)}{5/2 + \psi_0(0.5)}. \tag{8.52}$$

The expressions (8.45)–(8.49), (8.52) define the first-order approximations to the critical value (8.42) of the parameter μ that characterizes the rate of the chemical reaction and the corresponding canard (8.40), (8.41) of the system. This canard simulates the critical regime, separating slow chemical regimes from regimes with a self-acceleration.

Note the initial data for the system (8.37) are not fixed. With concrete initial data $x(0)$, $y(0)$, $z(0)$ we can glue the stable and unstable slow integral manifold at one point on the breakdown line (8.39). The canard passes through this point and corresponds to the initial value problem for (8.46), (8.48). Thus, a canard is a result of gluing stable and unstable slow integral manifolds at one point of the breakdown surface.

Let $\mu = \mu(y, \varepsilon)$ be given as a function. Then the gluing of the stable and unstable parts of slow integral manifolds can be realized at all points of the breakdown line (8.39) at the same time. This permits us to construct slow integral manifolds with changing stability (black swan) consisting entirely of canards. Each simulates the critical regime corresponding to the specified initial data and passes through a definite point on the breakdown line.

8.5.4.2 Black Swan Construction

We now take $\mu = \mu(y, \varepsilon)$ as control function. Then μ and the black swan $x = x(y, z, \varepsilon)$ have asymptotic expansions of the form:

$$\mu = \mu(y, \varepsilon) = \mu_0(y) + \varepsilon\mu_1(y) + \varepsilon^2\mu_2(y) + \dots ,$$
$$x = x(y, z, \varepsilon) = x_0(y, z) + \varepsilon x_1(y, z) + \varepsilon^2 x_2(y, z) + \dots .$$

Substituting these expansions into the invariance equation

$$\frac{\partial x}{\partial z}\frac{dz}{d\tau} + \frac{\partial x}{\partial y}\frac{dy}{d\tau} = \frac{dx}{d\tau},$$

which for the system (8.37) takes the form

$$\frac{\partial x(y, z, \varepsilon)}{\partial z}\varepsilon^{-1}(x(y, z, \varepsilon)z^2 + x(y, z, \varepsilon) - z) + \frac{\partial x(y, z, \varepsilon)}{\partial y}(z - y)$$
$$= \mu(y, \varepsilon)(5/2 + y) - x(y, z, \varepsilon)z^2 - x(y, z, \varepsilon),$$

and using the slow surface equation

$$x_0z^2 + x_0 - z = 0, \tag{8.53}$$

we obtain

$$\left(\frac{\partial x_0}{\partial z} + \varepsilon \frac{\partial x_1}{\partial z} + \varepsilon^2 \frac{\partial x_2}{\partial z} + \dots\right)(x_1 + \varepsilon x_2 + \dots)(1 + z^2)$$

$$+ \left(\frac{\partial x_0}{\partial y} + \varepsilon \frac{\partial x_1}{\partial y} + \varepsilon^2 \frac{\partial x_2}{\partial y} + \dots\right)(z - y)$$

$$= (\mu_0 + \varepsilon \mu_1 + \varepsilon^2 \mu_2 + \dots)(5/2 + y) - (x_0 + \varepsilon x_1 + \varepsilon^2 x_2 + \dots)z^2$$

$$- x_0 - \varepsilon x_1 - \varepsilon^2 x_2 - \dots . \tag{8.54}$$

Setting $\varepsilon = 0$ in (8.54) and taking (8.53) into account, we get

$$\frac{\partial x_0}{\partial z}(1 + z^2)x_1 = \mu_0(5/2 + y) - z. \tag{8.55}$$

Note that the relationship

$$\frac{\partial x_0}{\partial z} = \frac{1 - z^2}{(1 + z^2)^2} = 0$$

holds on the breakdown line (8.39). Noting this, by continuity of the function $x_1 = x_1(y, z)$, we require from (8.55) the condition

$$\mu_0 = \frac{1}{(5/2 + y)}$$

to ensure both sides of (8.55) are zero. From this, the expression for $\frac{\partial x_0}{\partial z}$, and (8.55) we have

$$x_1(y, z) = \frac{1 + z^2}{1 + z}. \tag{8.56}$$

Equating coefficients in ε in (8.54), we obtain

$$\frac{\partial x_1}{\partial y}(z - y) + \frac{\partial x_1}{\partial z}(1 + z^2)x_1 + \frac{\partial x_0}{\partial z}(1 + z^2)x_2 = \mu_1(5/2 + y) - x_1(1 + z^2). \tag{8.57}$$

To avoid a discontinuity in the function $x_2 = x_2(y, z)$ on the breakdown line we require the continuity condition:

$$\mu_1 = \frac{3}{(5/2 + y)}.$$

Applying (8.53), (8.56) and (8.57) yields

$$x_2(y, z) = \frac{[3(1 + z)^3 - 2z(2 + z)(1 + z^2)^2](1 + z^2)}{(1 + z)^3(1 - z^2)}.$$

Thus, we obtain the approximation to the black swan

$$x(y,z,\varepsilon) = \frac{z}{1+z^2} + \varepsilon\frac{1+z^2}{1+z} + \varepsilon^2\frac{\left[3(1+z)^3 - 2z(2+z)(1+z^2)^2\right](1+z^2)}{(1+z)^3(1-z^2)} + O(\varepsilon^3),$$

and the corresponding gluing function

$$\mu(y,\varepsilon) = \frac{\alpha(\varepsilon)}{(5/2+y)}, \quad \alpha(\varepsilon) = 1 + 3\varepsilon + O(\varepsilon^2). \tag{8.58}$$

For a given point $y = y^*$ on the breakdown line we can find the value $\mu^* = \mu(y^*,\varepsilon)$ from expression (8.58) which corresponds to the canard of the system. This trajectory lies on the black swan $x = x(y,z,\varepsilon)$ and passes through the point $y = y^*$ of the breakdown line. It should be noted that the choice of the gluing point $y = y^*$ is equivalent to the choice of the starting point of the trajectory, or the initial conditions.

Note that gluing the stable and unstable slow integral manifolds reduces the original system (8.37) to the following form

$$\frac{dx}{d\tau} = \alpha(\varepsilon) - xz^2 - x, \tag{8.59}$$

$$\varepsilon\frac{dz}{d\tau} = xz^2 + x - z, \tag{8.60}$$

$$\frac{dy}{d\tau} = z - y, \tag{8.61}$$

by the definition of $\mu(y,\varepsilon)$ in (8.58).

The system (8.59)–(8.61) has a black swan, which is a cylindrical surface. All trajectories on this surface are canards (see Figs. 8.13, 8.14, and 8.15), but only one of them is a limit cycle, and this cycle is asymptotically orbitally stable [158].

8.5.5   Gas Combustion in a Dust-Laden Medium

We now consider models of combustion of a rarefied gas mixture in an inert porous, or in a dusty, medium. We assume that the temperature distribution and phase-to-phase heat exchange are uniform. The chemical conversion kinetics are represented by a one-stage, irreversible reaction. The dimensionless model in this case has the form [56]

$$\varepsilon\dot{\theta} = \Psi(\eta)\exp\left(\theta/\left(1 + \beta\theta\right)\right) - \alpha(\theta - \theta_c) - \delta\theta,$$

$$\gamma_c\dot{\theta}_c = \alpha(\theta - \theta_c), \tag{8.62}$$

$$\dot{\eta} = \Psi(\eta)\exp\left(\theta/\left(1 + \beta\theta\right)\right),$$

$$\eta(0) = \eta_0/\left(1 + \eta_0\right) = \bar{\eta}_0, \quad \theta(0) = \theta_c(0) = 0.$$

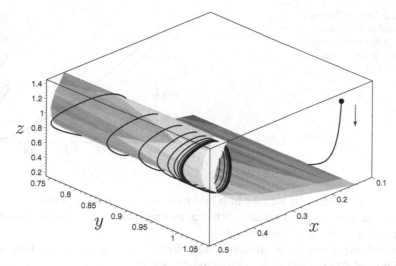

Fig. 8.13 The slow surface and the canard of the system (8.59)–(8.61) with $\varepsilon = 0.01$ and initial point $x = 0.1$, $y = 1$, $z = 1$. The trajectory of the system tends to the limit cycle (*dark line*)

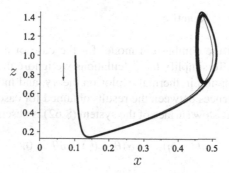

Fig. 8.14 xOz-projection of the canard of the system (8.59)–(8.61) with $\varepsilon = 0.01$ and initial point $x = 0.1$, $y = 1$, $z = 1$

Here, θ and θ_c are the dimensionless temperatures of the reactant phase and of the inert phase; η is the depth of conversion; η_0 is the parameter for autocatalyticity (this kinetic parameter characterizes the degree of self-acceleration of the reaction: the lower the value, the more marked the autocatalytic reaction will be); the small parameters β and ε characterize the physical properties of a gas mixture. The terms $-\delta\theta$ and $-\alpha(\theta - \theta_c)$ reflect the external heat dissipation and phase-to-phase heat exchange. The parameter γ_c characterizes the physical features of the inert phase. Depending on the relation between values of the parameters, the chemical reaction either moves to a slow regime with decay of the reaction, or into a regime of self-acceleration which leads to an explosion. So, if we change the value of one parameter, with fixed values of the other parameters, we can change the type of chemical reaction. Thus, it is possible to consider this problem as a special control

Fig. 8.15 xOy-projection of
the canard of the system
(8.59)–(8.61) with $\varepsilon = 0.01$
and initial point $x = 0.1$,
$y = 1, z = 1$

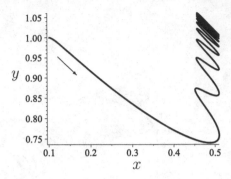

problem. For example, if we take heat loss from the gas phase as a control action,
we consider δ as a control variable. If the control variable is γ_c it means regulation
of the dust level in the reactant vessel.

The following two cases are considered: $\Psi(\eta) = 1 - \eta$ ($\eta_0 = 0$) (first-order
reaction) and $\Psi(\eta) = \eta(1 - \eta)$ (autocatalytic reaction).

8.5.5.1 Autocatalytic Reaction

Let us consider first the combustion model for the case of autocatalytic reaction
($\Psi(\eta) = \eta(1 - \eta)$). To simplify the calculations we ignore the small parameter β
(a widespread assumption in thermal explosion theory, and more detailed analysis
shows that the differences between the results obtained for cases $\beta = 0$ and $\beta \neq 0$
are not essential). The slow surface of the system (8.62) is described by the equation

$$\eta(1 - \eta)e^{\theta} - \alpha(\theta - \theta_c) - \delta\theta = 0. \tag{8.63}$$

The equation

$$\eta(1 - \eta)e^{\theta} - \alpha - \delta = 0, \tag{8.64}$$

(this is the derivative of (8.63) w.r.t. θ) together with (8.63) define the breakdown
curve $\mathscr{L}$, which separates the slow surface into stable and unstable parts.

We consider γ_c as a control parameter and recall that it means regulation of dust
levels in the reactant vessel. In this case we construct a special type of feedback
control.

We seek the critical function $\gamma_c(\theta, \varepsilon)$ and the black swan $\theta_c = \theta_c(\eta, \theta, \varepsilon)$ in the
form of asymptotic expansions:

$$\gamma_c = \Gamma_0(\theta) + \varepsilon\Gamma_1(\theta) + O(\varepsilon^2), \tag{8.65}$$

$$\theta_c = P_0(\eta, \theta) + \varepsilon P_1(\eta, \theta) + O(\varepsilon^2). \tag{8.66}$$

We use the black swan for the following reasons. We construct the canard modelling the critical regime with fixed initial point (or equivalently gluing point). However during a chemical process perturbations are possible. Due to the perturbations the trajectory of the system deviates from the canard and as a result a qualitative change of system behaviour is possible. When we construct the black swan, i.e. we glue the slow invariant manifold at all points of the breakdown curve, the trajectory of the system in the case of perturbations just goes from one canard to another. This means that there is no deviation from the selected regime.

From the invariance equations for system (8.62)

$$\varepsilon \gamma_c \frac{d\theta_c}{dt} = \gamma_c \left[\frac{\partial \theta_c}{\partial \theta} \varepsilon \frac{d\theta}{dt} + \varepsilon \frac{\partial \theta_c}{\partial \eta} \frac{d\eta}{dt} \right] \tag{8.67}$$

and asymptotic expansions (8.65), (8.66) we get

$$\left(\Gamma_0 + \varepsilon \Gamma_1 + \varepsilon^2 \Gamma_2 + \varepsilon^3 \ldots \right) \left[\frac{\partial P_0}{\partial \theta} + \varepsilon \frac{\partial P_1}{\partial \theta} + \varepsilon^2 \frac{\partial P_2}{\partial \theta} + \varepsilon^3 \ldots \right]$$
$$\times \left[\eta(1 - \eta)e^\theta - \alpha(\theta - P_0 - \varepsilon P_1 - \varepsilon^2 P_2 - \varepsilon^3 \ldots) - \delta\theta \right]$$
$$+ \varepsilon \left(\Gamma_0 + \varepsilon \Gamma_1 + \varepsilon^2 \Gamma_2 + \varepsilon^3 \ldots \right)$$
$$\times \left[\frac{\partial P_0}{\partial \eta} + \varepsilon \frac{\partial P_1}{\partial \eta} + \varepsilon^2 \frac{\partial P_2}{\partial \eta} + \varepsilon^3 \ldots \right] \eta(1 - \eta)e^\theta$$
$$= \varepsilon\alpha(\theta - P_0 - \varepsilon P_1 - \varepsilon^2 P_2 - \varepsilon^3 \ldots). \tag{8.68}$$

Equating coefficients in ε^0 and ε^1 in (8.68), we obtain

$$\eta(1 - \eta)e^\theta - \alpha[\theta - P_0(\eta, \theta)] - \delta\theta = 0,$$
$$\eta(1 - \eta)e^\theta \Gamma_0 \frac{\partial P_0}{\partial \eta} + \alpha P_1 \Gamma_0 \frac{\partial P_0}{\partial \theta} = \alpha(\theta - P_0),$$

which imply

$$P_0(\eta, \theta) = \alpha^{-1}[(\alpha + \delta)\theta - \eta(1 - \eta)e^\theta], \tag{8.69}$$
$$P_1(\eta, \theta) = \left[\alpha(\theta - P_0) - \eta(1 - \eta)e^\theta \frac{\partial P_0}{\partial \eta} \Gamma_0 \right] / \alpha \Gamma_0 \frac{\partial P_0}{\partial \theta}. \tag{8.70}$$

To ensure continuity of the function $P_1(\eta, \theta)$, since the relationship $\frac{\partial P_0}{\partial \theta} = 0$ holds on the breakdown line $\mathscr{L}$, see (8.63) and (8.64), from (8.70) we get

$$\left[\eta(1 - \eta)e^\theta \Gamma_0 \frac{\partial P_0}{\partial \eta} - \alpha(\theta - P_0) \right]_{\mathscr{L}} = 0,$$

or, using (8.64) and (8.69),

$$(\alpha + \delta)\Gamma_0 \sqrt{e^{2\theta} - 4(\alpha + \delta)e^{\theta}} = \alpha + \delta - \delta\theta.$$

Note that (8.64) gives us two expressions for η:

$$\eta = \frac{1 \pm \sqrt{1 - 4(\alpha + \delta)e^{-\theta}}}{2},$$

but we choose only one (with "+") because the function $\eta(\theta)$ monotonically increases. Finally we obtain

$$\Gamma_0(\theta) = \frac{\alpha(\alpha + \delta - \delta\theta)}{(\alpha + \delta)\sqrt{e^{2\theta} - 4(\alpha + \delta)e^{\theta}}}.$$

Similarly, equating the coefficients in ε^2 in (8.68) we obtain

$$\Gamma_1(\theta) = -\frac{\alpha P_1 + \Gamma_0\left[\eta(1-\eta)e^{\theta}\dfrac{\partial P_1}{\partial \eta} + \alpha P_1\dfrac{\partial P_1}{\partial \theta}\right]}{\eta(1-\eta)e^{\theta}\dfrac{\partial P_0}{\partial \eta}}.$$

Thus we have constructed the black swan that guarantees the safety of the physical process.

8.5.5.2 First-Order Reaction

The case of the first-order reaction ($\Psi(\eta) = (1 - \eta)$) is now studied. For simplicity we introduce the dimensionless concentration $\bar{\eta} = 1 - \eta$.

The slow surface of the system (8.62) is described by the equation

$$\bar{\eta}e^{\theta} - \alpha(\theta - \theta_c) - \delta\theta = 0. \tag{8.71}$$

The equation

$$\bar{\eta}e^{\theta} - \alpha - \delta = 0, \tag{8.72}$$

together with (8.71) define the breakdown curve $\mathscr{L}$, which separates the slow surface into stable and unstable parts.

We take $\gamma_c(\theta, \varepsilon)$ as control function. Then $\gamma_c(\theta, \varepsilon)$ and the black swan $\theta_c = \theta_c(\bar{\eta}, \theta, \varepsilon)$ have asymptotic expansions of the form:

$$\gamma_c = \Gamma_0(\theta) + \varepsilon\Gamma_1(\theta) + O(\varepsilon^2),$$

$$\theta_c = P_0(\bar{\eta}, \theta) + \varepsilon P_1(\bar{\eta}, \theta) + O(\varepsilon^2).$$

From the invariance equations (8.67) for system (8.62) and asymptotic expansions for γ_c and θ_c we get

$$\left(\Gamma_0 + \varepsilon\Gamma_1 + \varepsilon^2\Gamma_2 + \varepsilon^3 \dots\right)\left[\frac{\partial P_0}{\partial \theta} + \varepsilon\frac{\partial P_1}{\partial \theta} + \varepsilon^2\frac{\partial P_2}{\partial \theta} + \varepsilon^3 \dots\right]$$

$$\times \left[\bar{\eta}e^\theta - \alpha(\theta - P_0 - \varepsilon P_1 - \varepsilon^2 P_2 - \varepsilon^3 \dots) - \delta\theta\right]$$

$$-\varepsilon\left(\Gamma_0 + \varepsilon\Gamma_1 + \varepsilon^2\Gamma_2 + \varepsilon^3 \dots\right)\left[\frac{\partial P_0}{\partial \bar{\eta}} + \varepsilon\frac{\partial P_1}{\partial \bar{\eta}} + \varepsilon^2\frac{\partial P_2}{\partial \bar{\eta}} + \varepsilon^3 \dots\right]\bar{\eta}e^\theta$$

$$= \varepsilon\alpha\left[\theta - P_0 - \varepsilon P_1 - \varepsilon^2 P_2 - \varepsilon^3 \dots\right]. \tag{8.73}$$

Equating coefficients of ε^0 and ε^1 in (8.73), we obtain

$$\bar{\eta}e^\theta - \alpha(\theta - P_0) - \delta\theta = 0,$$

$$\alpha P_1 \Gamma_0 \frac{\partial P_0}{\partial \theta} - \bar{\eta}e^\theta \Gamma_0 \frac{\partial P_0}{\partial \bar{\eta}} = \alpha(\theta - P_0),$$

which imply

$$P_0(\bar{\eta}, \theta) = (\delta\theta - \bar{\eta}e^\theta)/\alpha + \theta, \tag{8.74}$$

$$P_1(\bar{\eta}, \theta) = \left[\alpha(\theta - P_0) + \bar{\eta}e^\theta\frac{\partial P_0}{\partial \bar{\eta}}\Gamma_0\right]\Big/\alpha\Gamma_0\frac{\partial P_0}{\partial \theta}. \tag{8.75}$$

To ensure continuity of the function $P_1(\bar{\eta}, \theta)$, since the relationship $\frac{\partial P_0}{\partial \theta} = 0$ holds on the breakdown line $\mathscr{L}$, from (8.75) we get

$$\Gamma_0(\theta) = \left[-\frac{\alpha(\theta - P_0)}{\bar{\eta}e^\theta\frac{\partial P_0}{\partial \bar{\eta}}}\right]\Big|_{\mathscr{L}},$$

or, using (8.72) and (8.74),

$$\Gamma_0(\theta) = \alpha\frac{\alpha + \delta - \delta\theta}{(\alpha + \delta)e^\theta}.$$

Then (8.75) takes the form

$$P_1(\bar{\eta}, \theta) = -\delta\theta\bar{\eta}e^\theta/(\alpha + \delta - \delta\theta).$$

Similarly we can find

$$\Gamma_1(\theta) = -\frac{\alpha^2\delta\theta\left[(\alpha + \delta - \delta\theta)(\alpha\delta\theta - \alpha - \delta) + \alpha\delta(\alpha + \delta)\right]}{(\alpha + \delta)^2(\alpha + \delta - \delta\theta)^2}.$$

Let us now consider the case when δ is a control function, which means that we control the combustion process by regulating the external heat dissipation [156].

By the change of variables:

$$\theta = \theta_0 + v, \quad \theta_0 = \ln \alpha \gamma_c^{-1},$$

$$\alpha a(v, \varepsilon) = \delta(\theta_0 + v, \varepsilon)(\theta_0 + v), \quad t = \alpha \gamma_c^{-1} \tau, \quad \bar{\eta} = 1 - \eta$$

the system (8.62) is transformed to the form:

$$\varepsilon \dot{v} = \bar{\eta} e^v - \gamma_c(\theta_0 + v - \theta_c) - \gamma_c a(v, \varepsilon),$$

$$\dot{\theta}_c = \theta_0 + v - \theta_c, \tag{8.76}$$

$$\dot{\bar{\eta}} = -\bar{\eta} e^v.$$

We find the critical function $a(v, \varepsilon)$ and slow integral manifold $\theta_c = \theta_c(\bar{\eta}, v, \varepsilon)$ in the form of asymptotic expansions:

$$\theta_c = P_0(\bar{\eta}, v) + \varepsilon P_1(\bar{\eta}, v) + \varepsilon^2 P_2(\bar{\eta}, v) + \varepsilon z,$$

$$a(v, \varepsilon) = a_0(v) + \varepsilon a_1(v) + \varepsilon^2 \omega, \tag{8.77}$$

when $P_0(\bar{\eta}, v)$ is defined by

$$\bar{\eta} e^v - \gamma_c(\theta_0 + v - P_0) - \gamma_c a_0 = 0. \tag{8.78}$$

Due to the condition $\frac{\partial P_0}{\partial v} = 0$ on the breakdown curve, from Eq. (8.78) we obtain

$$\bar{\eta} e^v - \gamma_c - \gamma_c \frac{\partial a_0}{\partial v} = 0. \tag{8.79}$$

The invariance equation

$$\varepsilon[\theta_0 + v - P_0 - \varepsilon P_1 - \varepsilon^2 P_2 - \varepsilon z] = \left[\frac{\partial P_0}{\partial v} + \varepsilon \frac{\partial P_1}{\partial v} + \varepsilon^2 \frac{\partial P_2}{\partial v} + \varepsilon \frac{\partial z}{\partial v}\right]$$

$$\times [\bar{\eta} e^v - \gamma_c(\theta_0 + v - P_0 - \varepsilon P_1 - \varepsilon^2 P_2 - \varepsilon z) - \gamma_c(a_0 + \varepsilon a_1 + \varepsilon^2 \omega)]$$

$$-\varepsilon\left[\frac{\partial P_0}{\partial \bar{\eta}} + \varepsilon \frac{\partial P_1}{\partial \bar{\eta}} + \varepsilon^2 \frac{\partial P_2}{\partial \bar{\eta}} + \varepsilon \frac{\partial z}{\partial \bar{\eta}}\right] \bar{\eta} e^v \tag{8.80}$$

follows from (8.76), (8.77).

Hence, for $z = 0$ and the terms at $O(\varepsilon)$ in (8.80) we obtain

$$\theta_0 + v - P_0 = \frac{\partial P_0}{\partial v} \gamma_c(P_1 - a_1) - \frac{\partial P_0}{\partial \bar{\eta}} \bar{\eta} e^v$$

or, due to (8.78),

$$\frac{\bar{\eta}e^v - \gamma_c a_0}{\gamma_c} = \gamma_c \frac{\partial P_0}{\partial v}(P_1 - a_1) - \frac{\partial P_0}{\partial \bar{\eta}}\bar{\eta}e^v. \tag{8.81}$$

Thus, on the breakdown curve we have:

$$\frac{\bar{\eta}e^v - \gamma_c a_0}{\gamma_c} = -\frac{\partial P_0}{\partial \bar{\eta}}\bar{\eta}e^v.$$

From this and (8.79) the equation for $a_0(v)$

$$\frac{\partial a_0}{\partial v}(1 - e^v) = a_0 - (1 - e^v) \tag{8.82}$$

follows.

The function $a_0(v)$:

$$a_0(v) = \frac{ve^v + 1 - e^v}{1 - e^v} \tag{8.83}$$

is the continuous solution of Eq. (8.82).

The function

$$P_1 = a_1 + \frac{\bar{\eta}e^v - \gamma_c a_0 - \bar{\eta}e^{2v}}{\gamma_c^2(1 + a_0\prime - \gamma_c^{-1}\bar{\eta}e^v)} \tag{8.84}$$

is determined from (8.78), (8.81).

The equation

$$-P_1 = \frac{\partial P_0}{\partial v}\gamma_c(P_2 - \omega) + \frac{\partial P_1}{\partial v}\gamma_c(P_1 - a_1) - \frac{\partial P_1}{\partial \bar{\eta}}\bar{\eta}e^v$$

is obtained from (8.80) with $z = 0$ at $O(\varepsilon^2)$.

Hence, on the breakdown curve we have

$$-P_1 = \frac{\partial P_1}{\partial v}\gamma_c(P_1 - a_1) - \frac{\partial P_1}{\partial \bar{\eta}}\bar{\eta}e^v.$$

From this and (8.83), (8.84) the equation for $a_1(v)$ is

$$\frac{\partial a_1}{\partial v}(1 - e^v) = a_1 - \frac{(1 - e^v)(1 + e^v)}{\gamma_c}.$$

The function

$$a_1(v) = \frac{1 - e^v}{\gamma_c}$$

is the continuous solution of this equation.

Thus, we obtain the control function

$$a(v, \varepsilon) = \frac{v e^v + 1 - e^v}{1 - e^v} + \varepsilon \frac{1 - e^v}{\gamma_c} + \varepsilon^2 \omega$$

or, in the original variables,

$$\delta = \delta(\theta, \varepsilon) = \theta^{-1} \left[\alpha \frac{\left(\theta - \ln \alpha \gamma_c^{-1} - 1 \right) e^\theta + \alpha \gamma_c^{-1}}{\alpha \gamma_c^{-1} - e^\theta} + \varepsilon(\alpha \gamma_c^{-1} - e^\theta) + O(\varepsilon^2) \right],$$

corresponding to the black swan $\theta_c = \theta_c(\eta, \theta, \varepsilon)$ of the system.

For a fixed point $\theta = \theta^*$ of the breakdown curve we can find the value δ^* from the last expression which corresponds to the canard of the system. This trajectory passes through the point θ^* of the breakdown curve and simulates the critical regime. It should be noted that the choice of the gluing point θ^* is equivalent to the choice of the initial point $\theta(0)$ of the trajectory. For example, with $\theta(0) = 0$, $\gamma_c = 1/6$, $\varepsilon = 0.01$, $\alpha = 2.34$ the critical regime corresponds to $\delta^* = 1.10797$. At $\delta < \delta^*$ we get a thermal explosion in the chemical system, at $\delta > \delta^*$ we get a non-explosive reaction.

Chapter 9
Appendix: Proofs

Abstract In this chapter we give the proof of a number of assertions from previous chapters. The level of mathematical sophistication required of the reader is higher than earlier chapters. For this reason, and if the readers' primary interest is in the application of the techniques, this chapter may be skipped.

9.1 The Existence and Properties of Bounded Solutions

In this book we have considered the differential system

$$\frac{dx}{dt} = f(x, y, t, \varepsilon),$$

$$\varepsilon \frac{dy}{dt} = g(x, y, t, \varepsilon),$$
(9.1)

and integral manifolds $y = h(x, t, \varepsilon)$ as a manifolds of solutions $x = x(t, \varepsilon), y = y(t, \varepsilon)$ which can be represented in the form

$$x = x(t, \varepsilon), \quad y = h(x(t, \varepsilon), t, \varepsilon).$$
(9.2)

If $h(x, t, \varepsilon)$ is a bounded function it is possible to say that the y-component of the class of solutions (9.2) is bounded or that these solutions are bounded with respect to the components of the vector y. We will now show how to prove the existence of an integral manifold for the simple case of solutions bounded on the real line. Moreover, to make the proof as simple as possible, we begin with the case of a scalar differential equation.

9.1.1 Scalar Linear Equation

We consider the scalar differential equation

$$\varepsilon \dot{y} = -ay + g(t),$$
(9.3)

© Springer International Publishing Switzerland 2014
E. Shchepakina et al., *Singular Perturbations*, Lecture Notes in Mathematics 2114,
DOI 10.1007/978-3-319-09570-7_9

with the positive coefficient a and a function $g(t)$ which is continuous and bounded on $(-\infty, \infty)$:

$$|g(t)| \leq M, \quad -\infty < t < \infty.$$

The function $\varphi(t, \varepsilon)$

$$\varphi(t, \varepsilon) = \frac{1}{\varepsilon} \int_{-\infty}^{t} \exp\left(-\frac{a(t-s)}{\varepsilon}\right) g(s) ds$$

is a solution to (9.3) which is defined and bounded on $(-\infty, \infty)$ because

$$|\varphi(t, \varepsilon)| \leq \left| \frac{1}{\varepsilon} \int_{-\infty}^{t} \exp\left(-\frac{a(t-s)}{\varepsilon}\right) |g(s)| ds \right| \leq \frac{1}{\varepsilon} \int_{-\infty}^{t} \exp\left(-\frac{a(t-s)}{\varepsilon}\right) M ds = \frac{M}{a},$$

and

$$\varepsilon \frac{d\varphi(t, \varepsilon)}{dt} = \frac{-a}{\varepsilon} \int_{-\infty}^{t} \exp\left(-\frac{a(t-s)}{\varepsilon}\right) g(s) ds + g(t) = -a\varphi(t, \varepsilon) + g(t).$$

We may consider $y = \varphi(t, \varepsilon)$ as a simple example of an integral manifold. We will show that $\varphi(t, \varepsilon)$ is attractive. Note that

$$\varphi(t_0, \varepsilon) = \frac{1}{\varepsilon} \int_{-\infty}^{t_0} \exp\left(-\frac{a(t-s)}{\varepsilon}\right) g(s) ds.$$

By linearity an arbitrary solution of (9.3) can be represented in the form

$$y(t, \varepsilon) = \exp\left(-\frac{a(t-t_0)}{\varepsilon}\right) y(t_0, \varepsilon) + \frac{1}{\varepsilon} \int_{t_0}^{t} \exp\left(-\frac{a(t-s)}{\varepsilon}\right) g(s) ds.$$

The same representation is true for $\varphi(t, \varepsilon)$, viz.,

$$\varphi(t, \varepsilon) = \exp\left(-\frac{a(t-t_0)}{\varepsilon}\right) \varphi(t_0, \varepsilon) + \frac{1}{\varepsilon} \int_{t_0}^{t} \exp\left(-\frac{a(t-s)}{\varepsilon}\right) g(s) ds.$$

Hence, for $y(t, \varepsilon) - \varphi(t, \varepsilon)$ we obtain the estimate

$$|y(t, \varepsilon) - \varphi(t, \varepsilon)| = \exp\left(-\frac{a(t - t_0)}{\varepsilon}\right) |y(t_0, \varepsilon) - \varphi(t_0, \varepsilon)|.$$

This proves the exponential stability, or attractivity, of $\varphi(t, \varepsilon)$.

9.1.2   Scalar Nonlinear Equation

Consider the following scalar differential equation

$$\varepsilon \dot{y} = -ay + g(y, t), \tag{9.4}$$

with positive coefficient a and function $g(y, t)$ which is continuous, bounded

$$|g(y, t)| \le M,$$

and satisfies the Lipschitz condition

$$|g(y, t) - g(\bar{y}, t)| \le L|y - \bar{y}|$$

on $t \in (-\infty, \infty)$, $y \in (-\infty, \infty)$, where L is a constant. If the function $\varphi(t, \varepsilon)$ is a solution to (9.4) which is defined and bounded on $(-\infty, \infty)$ then it satisfies the equation

$$\varphi(t, \varepsilon) = \frac{1}{\varepsilon} \int_{-\infty}^{t} \exp\left(-\frac{a(t - s)}{\varepsilon}\right) g(\varphi(s, \varepsilon), s) ds. \tag{9.5}$$

To check the solution (9.5) we consider the inequality

$$|\varphi(t, \varepsilon)| \le \left| \frac{1}{\varepsilon} \int_{-\infty}^{t} \exp\left(-\frac{a(t - s)}{\varepsilon}\right) |g(\varphi(s, \varepsilon), s)| ds \right|$$

$$\le \frac{1}{\varepsilon} \int_{-\infty}^{t} \exp\left(-\frac{a(t - s)}{\varepsilon}\right) M ds = \frac{M}{a},$$

and the identity

$$\varepsilon \frac{d\varphi(t, \varepsilon)}{dt} = \frac{-a}{\varepsilon} \int_{-\infty}^{t} \exp\left(-\frac{a(t - s)}{\varepsilon}\right) g(\varphi(s, \varepsilon), s) ds + g(\varphi(s, \varepsilon), t)$$

$$= -a\varphi(t, \varepsilon) + g(\varphi(t, \varepsilon), t).$$

Firstly we consider the case when L is small so that

$$\frac{L}{a} < 1. \tag{9.6}$$

In that case we introduce the metric space $C(D)$. Its elements are the scalar functions $p(t, \varepsilon)$, bounded by the value D i.e., $(|p(t, \varepsilon)| \leq D)$ and continuous for $t \in (-\infty, \infty)$ and $0 \leq \varepsilon \leq \varepsilon_0$ for sufficiently small positive ε_0. $C(D)$ is a complete space with the metric

$$d(p, \overline{p}) = \sup |p(t, \varepsilon) - \overline{p}(t, \varepsilon)|.$$

For the arbitrary $p, \overline{p} \in C(D)$ with $D \geq M/a$ we use (9.5) to construct the mapping $T(p)$:

$$T(p)(t, \varepsilon) = \frac{1}{\varepsilon} \int\limits_{-\infty}^{t} \exp\left(-\frac{a(t-s)}{\varepsilon}\right) g(p(s, \varepsilon), s) ds. \tag{9.7}$$

Then the inequality

$$|T(p)(t, \varepsilon)| \leq \frac{M}{a} \leq D,$$

follows immediately. This bound means that the operator $T(p)$ transforms the complete metric space $C(D)$ into itself.

The inequality

$$|T(p)(t, \varepsilon) - T(\overline{p})(t, \varepsilon)| \leq \left| \frac{1}{\varepsilon} \int\limits_{-\infty}^{t} \exp\left(-\frac{a(t-s)}{\varepsilon}\right) [g(p(s, \varepsilon), s) - g(\overline{p}(s, \varepsilon), s)] ds \right|$$

$$\leq \frac{1}{\varepsilon} \int\limits_{-\infty}^{t} \exp\left(-\frac{a(t-s)}{\varepsilon}\right) L |p(s, \varepsilon)) - \overline{p}(s, \varepsilon)| ds$$

$$\leq \frac{1}{\varepsilon} \int\limits_{-\infty}^{t} \exp\left(-\frac{a(t-s)}{\varepsilon}\right) L d(p, \overline{p}) ds = \frac{L}{a} d(p, \overline{p})$$

implies

$$d(T(p), T(\overline{p})) \leq \frac{L}{a} d(p, \overline{p})$$

with $L/a < 1$. This means that the operator $T(p)$ is contracting. Hence, due the Banach Contraction Principle $T(p)$ has a unique fixed point in $C(D)$. Thus Eq. (9.5), which may be rewritten as

$$p(t, \varepsilon) = T(p)(t, \varepsilon),$$

has a unique solution $p = \varphi(t, \varepsilon)$ in $C(D)$. To prove that this solution is attractive we need the following inequality [133]:

Gronwall-Bellman Inequality. Let I denote an interval on the real line of the form $[t_0, +\infty)$. Let c and β be nonnegative numbers and u be a real-valued continuous function defined on I. If u satisfies the integral inequality

$$u(t) \le c + \int_{t_0}^{t} \beta u(s) \, ds, \qquad \forall t \in I,$$

then

$$u(t) \le c \exp\left(\beta(t - t_0)\right), \qquad t \in I.$$

To check that $\varphi(t, \varepsilon)$ is attractive note that

$$\varphi(t_0, \varepsilon) = \frac{1}{\varepsilon} \int_{-\infty}^{t_0} \exp\left(-\frac{a(t - s)}{\varepsilon}\right) g(\varphi(s, \varepsilon), s) ds.$$

It is easily checked from (9.4) that

$$y(t, \varepsilon) = \exp\left(-\frac{a(t - t_0)}{\varepsilon}\right) y(t_0, \varepsilon) + \frac{1}{\varepsilon} \int_{t_0}^{t} \exp\left(-\frac{a(t - s)}{\varepsilon}\right) g(y(s, \varepsilon), s) ds,$$

is the solution that satisfies the arbitrary initial condition $y(t_0, \varepsilon)$. The same representation holds for $\varphi(t, \varepsilon)$:

$$\varphi(t, \varepsilon) = \exp\left(-\frac{a(t - t_0)}{\varepsilon}\right) \varphi(t_0, \varepsilon) + \frac{1}{\varepsilon} \int_{t_0}^{t} \exp\left(-\frac{a(t - s)}{\varepsilon}\right) g(\varphi(s, \varepsilon), s) ds.$$

Hence, for $y(t, \varepsilon) - \varphi(t, \varepsilon)$ we obtain the estimate

$$|y(t, \varepsilon) - \varphi(t, \varepsilon)| \le \exp\left(-\frac{a(t - t_0)}{\varepsilon}\right) |y(t_0, \varepsilon) - \varphi(t_0, \varepsilon)|$$

$$+ \frac{1}{\varepsilon} \int_{t_0}^{t} \exp\left(-\frac{a(t - s)}{\varepsilon}\right) |g(\varphi(s, \varepsilon), s) - g(y(s, \varepsilon), s)| ds$$

$$\le \exp\left(-\frac{a(t - t_0)}{\varepsilon}\right) |y(t_0, \varepsilon) - \varphi(t_0, \varepsilon)|$$

$$+ \frac{1}{\varepsilon} \int_{t_0}^{t} \exp\left(-\frac{a(t - s)}{\varepsilon}\right) L |\varphi(s, \varepsilon) - y(s, \varepsilon)| ds.$$

Multiplying the last inequality by

$$\exp\left(\frac{a(t-t_0)}{\varepsilon}\right),$$

we obtain the inequality

$$|y(t,\varepsilon)-\varphi(t,\varepsilon)|\exp\left(\frac{a(t-t_0)}{\varepsilon}\right)\le|y(t_0,\varepsilon)-\varphi(t_0,\varepsilon)|$$

$$+\frac{1}{\varepsilon}\int_{t_0}^{t}\exp\left(\frac{a(s-t_0)}{\varepsilon}\right)L|\varphi(s,\varepsilon)-y(s,\varepsilon)|ds.$$

Setting

$$c=|y(t_0,\varepsilon)-\varphi(t_0,\varepsilon)|,\quad \beta=\frac{L}{\varepsilon},\quad u=\exp\left(\frac{a(t-t_0)}{\varepsilon}\right)|\varphi(t,\varepsilon)-y(t,\varepsilon)|,$$

from the Gronwall–Bellman Inequality we obtain

$$|y(t,\varepsilon)-\varphi(t,\varepsilon)|\exp\left(\frac{a(t-t_0)}{\varepsilon}\right)\le|y(t_0,\varepsilon)-\varphi(t_0,\varepsilon)|\exp\left(\frac{L(t-t_0)}{\varepsilon}\right)$$

or

$$|y(t,\varepsilon)-\varphi(t,\varepsilon)|\le|y(t_0,\varepsilon)-\varphi(t_0,\varepsilon)|\exp\left(\frac{-(a-L)(t-t_0)}{\varepsilon}\right),$$

since $\frac{L}{a}\le 1$. This proves the exponential stability or attractivity of $\varphi(t,\varepsilon)$.

Thus, the existence and the attractivity of a slow invariant manifold are proved in the case $\mathbf{0+1}$ (see Sect. 3.1). Note, in conclusion, that the corresponding asymptotic expansion was also constructed there.

9.2   The Existence and Properties of Slow Integral Manifolds

We return now to the general case and consider again the non-autonomous system (9.1). For $\varepsilon=0$ we obtain the reduced system from (9.1), viz., $\frac{dx}{dt}=f(x,y,t,0)$, $0=g(x,y,t,0)$. Let the equation $g(x,y,t,0)=0$ have a solution $y=\phi(x,t)$, where the function $\phi(x,t)$ is defined for all $x\in\mathbb{R}^m, t\in\mathbb{R}$, and it is an isolated solution. The following conditions are satisfied:

(I) The functions f, g and ϕ are uniformly continuous and bounded together with their partial derivatives with respect to all variables up to the $(k + 2)$−order $(k \geq 0)$.
(II) The eigenvalues $\lambda_i(x, t)(i = 1, \ldots, n)$ of the matrix $B(x, t) = g_y(x, \phi(x, t), t, 0)$ satisfy the inequality

$$Re\lambda_i(x, t) \leq -2\gamma < 0. \tag{9.8}$$

These assumptions are extremely helpful in transforming the system (9.1), by the change of variables $y = y_1 + \phi(x, t)$, into the system

$$\dot{x} = X(x, y_1, t, \varepsilon),$$
$$\varepsilon\dot{y}_1 = B(x, t)y_1 + Y(x, y_1, t, \varepsilon), \tag{9.9}$$

where

$$B = g_y(x, \phi(x, t), t, 0),$$
$$X = f(x, y_1 + \phi(x, t), t, \varepsilon),$$
$$Y = g(x, y_1 + \phi(x, t), t, \varepsilon) - g_y(x, \phi(x, t), t, 0)y_1$$
$$-\varepsilon\phi_t(x, t) - \varepsilon\phi_x(x, t)X(x, y_1, t, \varepsilon).$$

Using the the Taylor-series expansion for a vector-function with the integral form of the remainder term we represent the function Y as follows:

$$Y = \int_0^1 [g_{yy}(x, \phi(x, t) + \theta y_1, t, 0)y_1]y_1 d\theta$$

$$+\varepsilon[\int_0^1 g_\varepsilon(x, \phi(x, t) + y_1, t, \theta\varepsilon)d\theta$$

$$-\phi_x(x, t)X(x, y_1, t, \varepsilon) - \phi_t(x, t)].$$

From (I) the following bounds may be obtained:

$$|X(x, y_1, t, \varepsilon)| \leq A, \tag{9.10}$$
$$\|Y(x, y_1, t, \varepsilon)\| \leq A(\|y_1\|^2 + \varepsilon), \tag{9.11}$$
$$\|B(x, t)\| \leq A, \tag{9.12}$$
$$\|X(x, y_1, t, \varepsilon) - X(\overline{x}, \overline{y}_1, t, \varepsilon)\| \leq A(\|x - \overline{x}\| + \|y_1 - \overline{y}_1\|), \tag{9.13}$$
$$\|Y(x, y_1, t, \varepsilon) - Y(\overline{x}, \overline{y}_1, t, \varepsilon)\| \leq A(\|\hat{y}_1\| + \varepsilon)(\|x - \overline{x}\| + \|y_1 - \overline{y}_1\|), \tag{9.14}$$
$$\|B(x, t) - B(\overline{x}, \overline{t})\| \leq A(\|x - \overline{x}\| + |t - \overline{t}|), \tag{9.15}$$

where A is a positive number and $\|\hat{y}_1\| = \max\{\|\overline{y}_1\|, \|y_1\|\}$. The inequalities (9.11)–(9.15) hold for all

$$-\infty < t, \overline{t} < \infty, x, \overline{x} \in \mathbb{R}^m, \|y_1\| \le \rho, \|\overline{y}_1\| \le \rho, 0 \le \varepsilon \le \varepsilon_0,$$

where ρ and ε_0 are positive numbers.
We shall denote by Ω the domain

$$\Omega = \{(x, y_1, t, \varepsilon) : t \in \mathbb{R}, \|y_1\| \le \rho, x \in \mathbb{R}^m, 0 \le \varepsilon \le \varepsilon_0\},$$

where $X(x, y_1, t, \varepsilon)$, $B(x, t)$ and $Y(x, y_1, t, \varepsilon)$ are defined above.

Our attention will be focused on the slow integral manifolds of the system (9.9) which are described by the following equations:

$$y_1 = p(x, t, \varepsilon). \tag{9.16}$$

We shall assume that function p is defined in the domain

$$\Omega_1 = \{(x, t, \varepsilon) : x \in \mathbb{R}^m, t \in \mathbb{R}, 0 \le \varepsilon \le \varepsilon_0\},$$

is continuous with respect to t and ε in this domain, satisfies the Lipschitz condition with respect to x and a constant independent of t:

$$\|p(x, t, \varepsilon) - p(\overline{x}, t, \varepsilon)\| \le \Delta \|x - \overline{x}\|, \tag{9.17}$$

and its norm is bounded

$$\|p(x, t, \varepsilon)\| \le \Delta, \tag{9.18}$$

for some $\Delta > 0$. If the trajectory $(x(t), y_1(t), t)$ belongs to the integral manifold (9.16), then $y_1(t) = p(x(t), t, \varepsilon)$. Functions $x(t)$ and $y_1(t) = p(x(t), t, \varepsilon)$ must satisfy the system (9.9), and then

$$\dot{x} = X(x, p(x, t, \varepsilon), t, \varepsilon). \tag{9.19}$$

From (9.13) and (9.17) we obtain the inequality

$$\|X(x, p(x, t, \varepsilon), t, \varepsilon) - X(\overline{x}, p(\overline{x}, t, \varepsilon), t, \varepsilon)\| \le A(1 + \Delta)\|x - \overline{x}\|. \tag{9.20}$$

Hence the norm of the right side of Eq. (9.19), $\|X\|$, is bounded (with the constant A) and satisfies the Lipschitz condition with the constant $A(1 + \Delta)$ independently of t. Thus there is a unique solution $x = \varphi(t)$ to Eq. (9.19) at every $x_0 \in \mathbb{R}^m$ with an initial condition at $t = \tau$:

$$\varphi(t) = \Phi(t, \tau, x_0, \varepsilon \mid p), \quad \Phi(\tau, \tau, x_0, \varepsilon \mid p) = x_0,$$

defined at $t \in \mathbb{R}$. Here the notation $(\cdots \mid p)$ underlines the dependence of the solution on the function p.

The function $y_1 = p(\Phi(t), t, \varepsilon)$ is the bounded solution of the equation

$$\varepsilon \dot{y}_1 = B(\varphi(t), t) y_1 + Y(\varphi(t), y_1, t, \varepsilon). \tag{9.21}$$

for all $t \in \mathbb{R}$. For this reason it must satisfy the integral equation [29] [compare with (9.5)]

$$y(\tau) = \varepsilon^{-1} \int_{-\infty}^{\tau} U_\varphi(\tau, t, \varepsilon) Y(\varphi(t), y(t), t, \varepsilon) dt, \tag{9.22}$$

where $U_\varphi(t, s, \varepsilon)$ is the fundamental matrix of the homogeneous equation

$$\varepsilon \dot{y} = B(\varphi(t), t) y, \quad U_\varphi(s, s, \varepsilon) = I.$$

The following bound, which provides the convergence of the improper integral in (9.22), will be established below:

$$\|U_\varphi(\tau, t, \varepsilon)\| \le K \varepsilon^{-\frac{\gamma}{\varepsilon}(\tau - t)} \quad K \ge 1, -\infty < t \le \tau < \infty. \tag{9.23}$$

Let $x_0 = x$ and $\varphi(t) = \Phi(t, \tau, x, \varepsilon \mid p)$. Then from (9.22) we get the following equation for the function $p(x, t, \varepsilon)$:

$$p(x, \tau, \varepsilon) = \varepsilon^{-1} \int_{-\infty}^{\tau} U_\varphi(\tau, t, \varepsilon) Y(\varphi(t), p(\varphi(t), t, \varepsilon), t, \varepsilon) dt. \tag{9.24}$$

On the other hand, if Eq. (9.24) has a solution which satisfies (9.17), (9.18), then (9.24) defines the slow integral manifold of system (9.9). We shall give a brief justification of this fact. For an arbitrary point (x_0, y_{10}, t_0), belonging to the surface $y_1 = p(x, t, \varepsilon)$, (that is, satisfying the relation $y_{10} = p(x_0, t_0, \varepsilon)$) Eq. (9.19) has a solution $x = \varphi(t) = \Phi(t, t_0, x_0, \varepsilon \mid p))$. Note that Eq. (9.24) implies $t < \tau$. The equality (9.24) and

$$\Phi(t, t_0, x_0, \varepsilon \mid p) = \Phi(t, \tau, \Phi(\tau, t_0, x_0, \varepsilon \mid p), \varepsilon \mid p)$$

yield the result that the function $y_1 = p(\varphi(t), t, \varepsilon)$ is the solution of the Eq. (9.21). This equality says that the solution at time t, with initial point (x_0, t_0), may be reached by first going to the time τ and using this as the initial time for proceeding to t. Hence Eq. (9.24) may be considered as an operator equation for p.

Now we obtain some auxiliary inequalities. The following statement will often be used (see, for example, [133]).

Theorem 3 (Integral Inequality). *Let $u(t)$ be a continuous and positive function (for all $t \in [t_0, t_0 + T]$), which satisfies the following inequality:*

$$u(t) \leq f(t) + \int_{t_0}^{t} [\varphi_1(t)\varphi_2(s)u(s) + \psi(t, s)]ds,$$

where $f(t), \varphi_1(t), \varphi_2(t), \psi(t, s)$ are continuous, nonnegative functions for all $t \in [t_0, t_0 + T]$. Then the following inequality holds:

$$u(t) \leq u_0(t) = f(t) + \int_{t_0}^{t} \psi(t, s)ds + \varphi_1(t) \int_{t_0}^{t} \varphi_2(\tau) f(\tau) \exp\left(\int_{\tau}^{t} \varphi_1(s)\varphi_2(s)ds\right) d\tau$$

$$+ \varphi_1(t) \int_{t_0}^{t} \exp\left(\int_{\tau}^{t} \varphi_1(s)\varphi_2(s)ds\right) \varphi_2(\tau) \left(\int_{t_0}^{\tau} \psi(t, s)ds\right) d\tau.$$

We note that T may be arbitrarily large. Hence, this estimate is valid for all $t \geq t_0$ if the functions f, Φ_1, Φ_2, ψ are defined for $t_0 \leq s \leq t < \infty$. Similarly, the bound

$$u(t) \leq f(t) + \int_{t}^{t_0} [\varphi_1(t)\varphi_1(s)u(s) + \psi(t, s)]ds$$

may be considered for $t \leq t_0$, which yields the inequality $u(t) \leq u_0(t)$.

We introduce the metric space $C(D, \Delta)$. Its elements are the functions $p(x, t, \varepsilon)$, bounded and continuous in Ω_1, with their values in $\mathbb{R}^n$. Besides, these functions satisfy the conditions (9.17), (9.18) with the metric

$$d(p, \overline{p}) = \sup_{\Omega_1} \| p(x, t, \varepsilon) - \overline{p}(x, t, \varepsilon)\|.$$

For arbitrary $p, \overline{p} \in C(D, \Delta)$ we consider Eq. (9.19) and establish the validity of the following statement.

Lemma 1. *Let $A(1 + \Delta) \leq \alpha$. Then the inequality*

$$\|\Phi(t, \tau, x, \varepsilon \mid p) - \Phi(t, \tau, \overline{x}, \varepsilon \mid \overline{p})\| \leq \|x - \overline{x}\|e^{\alpha(\tau - t)} + \frac{d(\overline{p}, p)}{1 + \Delta}(e^{\alpha(\tau - t)} - 1)$$

$$(9.25)$$

holds for all $\tau \geq t$.

Proof. It should be noted that the functions $\varphi(t) = \Phi(t, \tau, x, \varepsilon \mid p)$ and $\overline{\varphi}(t) = \Phi(t, \tau, x, \varepsilon \mid p)$ satisfy the integral equations

$$\varphi(t) = x + \int_{\tau}^{t} X\left(\varphi(s), p(\varphi(s), s, \varepsilon), s, \varepsilon\right) ds$$

and

$$\overline{\varphi}(t) = \overline{x} + \int_{\tau}^{t} X\left(\overline{\varphi}(s), \overline{p}(\overline{\varphi}(s), s, \varepsilon), s, \varepsilon\right) ds.$$

as solutions of corresponding differential equations with initial conditions $\varphi(\tau) = x$ and $\overline{\varphi}(\tau) = \overline{x}$.

From (9.14), (9.17) and (9.18), taking into account

$$\|p(\varphi(s), s, \varepsilon), s, \varepsilon) - \overline{p}(\overline{\varphi}(s), s, \varepsilon)\| \le \|p(\varphi(s), s, \varepsilon), s, \varepsilon) - p(\overline{\varphi}(s), s, \varepsilon)\|$$
$$+ \|p(\overline{\varphi}(s), s, \varepsilon), s, \varepsilon) - \overline{p}(\overline{\varphi}(s), s, \varepsilon)\|$$
$$\le \Delta \|\varphi(s) - \overline{\varphi}(s)\| + d(p, \overline{p}),$$

we obtain, for $\tau \ge t$

$$\|\varphi(t) - \overline{\varphi}(t)\| \le \|x - \overline{x}\| + \int_{t}^{\tau} \|X\left(\varphi(s), p(\varphi(s), s, \varepsilon), s, \varepsilon\right)$$

$$-X\left(\overline{\varphi}(s), \overline{p}(\overline{\varphi}(s), s, \varepsilon), s, \varepsilon\right)\| ds$$

$$\le \|x - \overline{x}\| + \int_{t}^{\tau} A[(1 + \Delta)\|\varphi(s) - \overline{\varphi}(s)\| + \|p - \overline{p}\|] ds$$

$$\le \|x - \overline{x}\| + \int_{t}^{\tau} A[(1 + \Delta)\|\varphi(s) - \overline{\varphi}(s)\| + d(p, \overline{p})] ds.$$

Using the Theorem 3 we derive the desired estimate (9.25), on letting

$$\|x - \overline{x}\| = f(t), \quad A(1 + \Delta) = \varphi_1(t), \quad \varphi_2(s) = 1, \quad Ad(p, \overline{p}) = \psi(t, s), \quad \tau \ge t_0$$

and the right-hand side of (9.25) plays the role of $u_0(t)$.   □

We need the following statement to justify the inequality (9.23) [197]:

Lemma 2. *Let the matrix* $A(t)(-\infty < t < \infty)$ *be bounded and satisfy the Lipschitz condition with respect to* t *with the constant* q. *If the real parts of the eigenvalues of matrix* $A(t)$ *do not exceed the number* -2γ ($\gamma > 0$) *for all* t, *then there exist positive numbers* K *and* ε_0, *such that the fundamental matrix* $U(t, s, \varepsilon)$, $U(s, s, \varepsilon) = I$, *of the equation*

$$\varepsilon \dot{z} = A(t)z,$$

admits the bound

$$\|U(t,s,\varepsilon)\| \le K e^{-\gamma(t-s)/\varepsilon},$$

for all $-\infty < s \le t < \infty, 0 < \varepsilon \le \varepsilon_0$.

To prove estimate (9.23) it is sufficient to point out from (9.12) and (9.15), that $\|B(t,\varphi(t))\| \le A$ and $\|B(\varphi(t),t) - B(\varphi(\bar{t}),\bar{t})\| \le A(|t - \bar{t}|) + \|\varphi(t) - \varphi(\bar{t})\| \le A(1 + A)|t - \bar{t}|$, since $\|\varphi(t) - \varphi(\bar{t})\| \le A|t - \bar{t}|$.

This means that the norm of matrix $B(\varphi(t),t)$ is bounded and satisfies the Lipschitz condition with respect to t, for all real t. Hence, the matrix $A(t) = B(\varphi(t),t)$ satisfies the conditions of Lemma 2, and the inequality (9.23) follows.

Obviously, the matrix $U_\varphi(\tau,t,\varepsilon)$ depends on the choice of the function φ. We shall evaluate the norm of the difference $U_\varphi(\tau,t,\varepsilon) - U_{\overline{\varphi}}(\tau,t,\varepsilon)$, where $\varphi = \Phi(t,\tau,x,\varepsilon \mid p), \overline{\varphi} = \Phi(t,\tau,\overline{x},\varepsilon \mid \overline{p})$.

The fundamental matrices U_φ and $U_{\overline{\varphi}}$ are given by the following differential equations with initial conditions:

$$\varepsilon \frac{dU_\varphi}{d\tau} = B(\varphi(\tau))U_\varphi, U_\varphi(t,t,\varepsilon) = I;$$

$$\varepsilon \frac{dU_{\overline{\varphi}}}{d\tau} = B(\overline{\varphi}(\tau))U_{\overline{\varphi}}, U_{\overline{\varphi}}(t,t,\varepsilon)) = I.$$

Subtracting the second equality from the first

$$\varepsilon\left(\frac{d(U_\varphi - U_{\overline{\varphi}})}{d\tau}\right) = B(\varphi(\tau))\left(U_\varphi - U_{\overline{\varphi}}\right)$$

$$+ (B(\varphi(\tau)) - B(\overline{\varphi}(\tau))) U_{\overline{\varphi}}, U_\varphi(t,t,\varepsilon) - U_{\overline{\varphi}}(t,t,\varepsilon)) = 0$$

one can represent the result in integral form

$$U_\varphi(\tau,t,\varepsilon) - U_{\overline{\varphi}}(\tau,t,\varepsilon) = \varepsilon^{-1} \int_t^\tau U_\varphi(\tau,s,\varepsilon)[B(\varphi(s),s) - B(\overline{\varphi}(s),s)]U_{\overline{\varphi}}(s,t,\varepsilon)ds.$$

The estimates (9.15) and (9.23) yield, for $\tau \ge t$,

$$\|U_\varphi(\tau,t,\varepsilon) - U_{\overline{\varphi}}(\tau,t,\varepsilon)\| \le K^2 A \varepsilon^{-1} e^{-\gamma\varepsilon^{-1}(\tau-t)} \int_t^\tau \|\varphi(s) - \overline{\varphi}(s)\|ds. \quad (9.26)$$

We apply inequality (9.25) and obtain from (9.26), for $\varepsilon\alpha \le \gamma/2$, the inequality

$$\|U_\varphi(\tau,t,\varepsilon) - U_{\overline{\varphi}}(\tau,t,\varepsilon)\| \leq \frac{2K^2A}{\gamma}e^{-\gamma(\tau-t)/2\varepsilon}[\|x - \overline{x}\| + \frac{d(p,\overline{p})}{1+\Delta}]. \quad (9.27)$$

If we introduce the mapping T:

$$T(p)(\tau,x) = \varepsilon^{-1}\int_{-\infty}^{\tau} U_\phi(\tau,t,\varepsilon)Y(\phi(t), p(\phi(t),t,\varepsilon),t,\varepsilon)dt, \quad (9.28)$$

where $\varphi(t) = \Phi(t,\tau,x,\varepsilon \mid p)$, then the following statement is valid.

Lemma 3. *The mapping $T(p)(\tau,x)$ satisfies the following inequalities:*

$$\|T(p)(\tau,x)\| \leq KA\gamma_{-1}(D_2 + \varepsilon); \quad (9.29)$$

$$\|T(p)(\tau,x) - T(p)(\tau,\overline{x})\| \leq \frac{2KA}{\gamma}[(D + \varepsilon)(1 + \Delta)$$

$$+ \frac{2KA}{\gamma}(D_2 + \varepsilon)]\|x - \overline{x}\|; \quad (9.30)$$

$$\|T(p)(\tau,x) - T(\overline{p})(\tau,x)\| \leq \frac{2KA}{\gamma(1+\Delta)}[(D + \varepsilon)(1 + \Delta)$$

$$+ \frac{2KA}{\gamma}(D_2 + \varepsilon)]d(p,\overline{p}). \quad (9.31)$$

Proof. The inequalities (9.11), (9.18), (9.23) yield the following estimate

$$\|T(p)(\tau,x)\| \leq \varepsilon^{-1}\int_{-\infty}^{\tau} Ke^{-\gamma\varepsilon^{-1}(\tau-t)}A(D^2 + \varepsilon)dt = KA\gamma^{-1}(D^2 + \varepsilon).$$

Using the bounds (9.14), (9.17), (9.23), (9.25) and (9.27) we obtain

$$\|T(p)(\tau,x) - T(\overline{p})(\tau,\overline{x})\| \leq \varepsilon^{-1}\int_{-\infty}^{\tau} [\|U_\phi(\tau,t,\varepsilon)\| \times \|Y(\phi(t), p(\phi(t),t,\varepsilon),t,\varepsilon)$$

$$- Y(\overline{\phi}(t), \overline{p}(\overline{\phi}(t),t,\varepsilon),t,\varepsilon)\| + \|U_\phi(\tau,t,\varepsilon)$$

$$- U_{\overline{\phi}}(\tau,t,\varepsilon)\| \cdot \|Y(\overline{\phi}(t), \overline{p}(\overline{\phi}(t),t,\varepsilon),t,\varepsilon)\|]dt$$

$$\leq \varepsilon^{-1}\int_{-\infty}^{\tau} \{Ke^{-\gamma\varepsilon^{-1}(\tau-t)}A(D + \varepsilon)[(1 + \Delta)\|\phi(t)$$

$$- \overline{\phi}(t)\| + d(p,\overline{p})] + \frac{2K^2A}{\gamma}[\|x - \overline{x}\|$$

$$+ \frac{d(p,\overline{p})}{1+\Delta}]A(D^2 + \varepsilon)e^{-\gamma(\tau-t)/(2\varepsilon)}\}dt$$

and, finally,

$$\|T(p)(\tau, x) - T(\overline{p})(\tau, \overline{x})\| \le \frac{2KA}{\gamma}[(D + \varepsilon)(1 + \varDelta)$$

$$+ \frac{2KA}{\gamma}(D^2 + \varepsilon)][\|x - \overline{x}\| + \frac{d(p, \overline{p})}{1 + \varDelta}]. \quad (9.32)$$

Setting $p = \overline{p}$ and $x = \overline{x}$ one at a time in the last inequality we obtain the required bounds (9.30), (9.31).

The proof is now complete.   □

Assume now that $D = \varepsilon D_0$, $\varDelta = \varepsilon \varDelta_0$, while the set $C(D, \varDelta)$ is constructed.

The numbers D_0 and $\varDelta_0$ will be chosen to obtain the following inequalities for sufficiently small ε $(0 < \varepsilon \le \varepsilon_0)$:

$$2\varepsilon A(1 + \varepsilon \varDelta_0) \le \gamma, \tag{9.33}$$

$$KA\gamma^{-1}(1 + \varepsilon D_0^2) \le D_0, \tag{9.34}$$

$$\frac{2KA}{\gamma}[(D_0 + 1)(1 + \varepsilon \varDelta_0) + \frac{2KA}{\gamma}(1 + \varepsilon D_0^2)] \le \varDelta_0, \tag{9.35}$$

$$\varepsilon \varDelta_0/(1 + \varepsilon \varDelta_0) < 1. \tag{9.36}$$

Inequalities (9.30) and (9.31) imply

$$\|T(p)(\tau, x)\| \le \varepsilon D_0,$$

$$\|T(p)(\tau, x) - T(p)(\tau, \overline{x})\| \le \varepsilon \varDelta_0 \|x - \overline{x}\|.$$

These bounds mean that the operator $T(p)$ transforms the complete metric space $C(\varepsilon D_0, \varepsilon \varDelta)$ into itself.

We use the exact upper bound in (9.32) with respect to t and x. Then (9.36) yields the existence of a positive number $q < 1$, such that

$$d(T(p), T(\overline{p})) \le qd(p, \overline{p}).$$

This means that the operator $T(p)$ is contracting. Hence it has a unique fixed point in $C(\varepsilon D_0, \varepsilon \varDelta_0)$. Thus the Eq. (9.24), which may be rewritten as

$$p(x, \tau, \varepsilon) = T(p)(\tau, x),$$

has a unique solution $p^*(x, t, \varepsilon)$ in $C(\varepsilon D_0, \varepsilon \varDelta_0)$. Consequently, the system (9.9) has an integral manifold $y = p^*(x, t, \varepsilon)$. It may be noted that the system (9.9) was obtained from (2.1) by the change of variables $y = y_1 + \phi(x, t)$. Hence, the system (2.1) has the integral manifold $y = h(x, t, \varepsilon) = \phi(x, t) + p^*(x, t, \varepsilon)$. The above argument permits us to formulate the following statement.

Theorem 4. *Let the conditions (I),(II) at the beginning section 9.2 hold. Then there exist $\varepsilon_1(0 < \varepsilon_1 \leq \varepsilon_0)$ such that for $\varepsilon \in (0, \varepsilon_1)$ the system (9.1) has an integral manifold of slow motions $y = h(x, t, \varepsilon)$. The motion along this manifold is described by the equation*

$$\dot{x} = f(x, h(x, t, \varepsilon), t, \varepsilon). \tag{9.37}$$

Remark 9.1. If $f(0, 0, t, \varepsilon) \equiv 0, g(0, 0, t, \varepsilon) \equiv 0$ hold then $h(0, t, \varepsilon) \equiv 0$.

We shall now distinguish some essential properties of the slow integral manifolds.

The first is connected with the smoothness of the integral manifolds: under the conditions of Theorem 3 the function $h(x, t, \varepsilon)$ has bounded partial derivatives with respect to x and t up to and including the k-order.

In many applications there are autonomous, periodic or almost-periodic differential systems. Hence the question arises as to whether the integral manifold has the same properties.

The following is a fact. If the functions f and g do not depend on t, or are periodic or almost-periodic with respect to t, then function $h(x, t, \varepsilon)$ has the same properties, see [197].

9.3   Justification of Asymptotic Representation

To justify the asymptotic expansion of $h(x, t, \varepsilon)$ expansion we may follow the same scheme, as in the proof of the existence of integral manifold $y_1 = p^*(x, t, \varepsilon)$. We shall denote by $p_k(x, t, \varepsilon)$ the finite sum

$$p_k = \sum_{i=0}^{k} \varepsilon^i h_i(x, t)$$

where h_i are the coefficients of the expansion computed according to (2.24). We make the change of variables $y = y_1 + p_k(x, t, \varepsilon)$ in the system (2.1). Then for the variables x and y we obtain the system of the form (9.9), where the functions X and Y are as follows:

$$X = f(x, y_1 + p_k(x, t, \varepsilon), t, \varepsilon)$$

$$Y = g(x, y_1 + p_k(x, t, \varepsilon), t, \varepsilon) - g_y(x, \phi(x, t), t, 0)y_1$$

$$-\varepsilon \frac{\partial p_k}{\partial t}(x, t, \varepsilon) - \varepsilon \frac{\partial p_k}{\partial x}(x, t, \varepsilon)X(x, y_1, t, \varepsilon).$$

It should be noted that the functions X and Y satisfy inequalities analogous to (9.10), (9.11), (9.13), (9.14), where the inequality (9.11) must be replaced by the bound

$$\|Y\| \le A_0(\|y_1\|^2 + \varepsilon\|y_1\| + \varepsilon^{k+1}).$$

Then the operator $T(p)(\tau, x)$ for y should be considered on the set $C(\varepsilon^{k+1}D_k, \varepsilon^{k+1}\Delta_k)$, where D_k is a positive number with the restriction

$$\frac{KA_0}{\gamma}[(\varepsilon^{k+1}D_k)^2 + \varepsilon(\varepsilon^{k+1}D_k) + \varepsilon^{k+1}] \le \varepsilon^{k+1}D_k.$$

Thus we obtain the existence of the integral manifold $y_1 = p_{k+1}^*(x, t, \varepsilon)$, where $\|p_{k+1}^*(x, t, \varepsilon)\| \le \varepsilon^{k+1}D_k$. This means that system (2.1) has an integral manifold, which may be represented as follows

$$y = \phi(x, t) + \varepsilon h_1(x, t) + \cdots + \varepsilon^k h_k(x, t) + \varepsilon^{k+1}h_{k+1}(x, t, \varepsilon),$$

where $\varepsilon^{k+1}h_{k+1} = p_{k+1}^*$ and h_{k+1} is a smooth function with bounded norm.

Bibliographical Remarks

The origins of the method of integral manifolds are found in the works of J. Hadamard [70], A. Lyapunov [100], H. Poincare [139] and O. Perron [136]. The essence of the method of integral manifolds was realized with amazing depth by A. Lyapunov [100] who used order reduction when he investigated the critical cases of one zero, or a pair of purely imaginary, eigenvalues. The possibility of lowering the dimensionality of the system is the essential aspect of the method of integral manifolds. The foundations of the theory were laid by N. Bogolyubov [13] and significant impact on the development of the method was provided by N. Bogolyubov and Yu Mitropolskii [14, 15] and J. Hale [71–73].

As to the singularly perturbed systems, pioneering papers were published during 1957–1970 by such as K. Zadiraka, V. Fodchuk and Ya. Baris from the scientific school headed by N. Bogolyubov and Yu. Mitropolsky in the Institute of Mathematics of Ukrainian Academy of Science, Kiev. The existence of slow integral manifolds, stable [217, 218], unstable and conditionally stable [4, 5], are shown in these papers. Some of these were translated into English on the initiative of Jack Hale and AMS, the main results of [217] can be found in [112]. The authors of [53, 90] gave a short but realistic description of the history of the geometrical theory of singular perturbations. At the same time a series of papers devoted to the existence and asymptotic expansions of integral manifolds for nonautonomous differential systems with slow and fast variables were published by Yu Mitropolskii and O. Lykova. These results can be found in the books [113, 114], see also the books [29, 185, 197, 210]. Various aspects of the theory of slow integral manifolds and the behavior of solutions in their neighborhood are presented in [27, 38, 45, 58, 74, 75, 81, 92, 105, 117, 120, 126, 191, 200], see also references therein.

The concept of a fast integral manifold was introduced in [170]. This allowed the construction of a smooth transformation which reduced the original singularly perturbed differential system to a block-diagonal form. This means that the system under consideration is decomposed into two subsystems, the first of which is independent and regularly perturbed with respect to ε and the second one describes

© Springer International Publishing Switzerland 2014

E. Shchepakina et al., *Singular Perturbations*, Lecture Notes in Mathematics 2114, DOI 10.1007/978-3-319-09570-7

fast components of solutions. Some theoretical and applied results along these lines were obtained for ODE's [167, 168, 171–173, 175, 176, 178, 206, 208], for PDE's [12, 174, 177, 181], non-Lipschitzian [147], discontinuous [182], discrete [207, 209] and difference-differential systems [48–50].

In the first papers devoted to canards non-standard analysis was the main tool of investigations [7, 8, 16, 35, 36, 221], matched asymptotics were used in [42, 110], the Gevrey version of matched asymptotic expansions (see references in [51]), the approach based on the blow-up technique in [39], and on the technique of upper and lower solutions in [26, 123].

In many papers devoted to canards the term "canard" is associated with periodic trajectories [7, 8, 20, 21, 35, 110]. In the papers [59, 60] it was suggested a canard is a one-dimensional slow invariant manifold if it contains a stable slow invariant manifold and an unstable one, and a canard is obtained as a result of gluing stable (attractive) and unstable (repulsive) slow invariant manifolds at one point of the breakdown surface due to the availability of an additional scalar parameter which may be considered as a control parameter. This approach was proposed for the first time in [59, 60] and was then applied to construct canards in $\mathbb{R}^3$ [56, 61, 184], canards for PDE [60, 61] and canard travelling waves [149, 189]. Moreover, the use of control functions instead of control parameters allowed the construction of black swans [61, 155–158, 161–164, 184, 188], and canard cascades [180], the consideration of the effect of delayed loss of stability [124, 125] and [150, 157, 162], and the solution of a number of applied problems [56, 61, 156, 159, 160, 183, 184]. Different kinds of canards, canards in piecewise linear systems, the influence of stochastic perturbations, mixed mode oscillations and miscellaneous applications were considered in [10, 11, 18–20, 23, 24, 30–32, 34, 40, 47, 63, 64, 67, 68, 79, 82, 104, 115, 135, 137, 140–144, 151, 154, 198, 212–215, 220], see also the overview [33].

References

1. Adegbie, K.S., Alao, F.I.: Studies on the effects of interphase heat exchange during thermal explosion in a combustible dusty gas with general Arrhenius reaction-rate laws. J. Appl. Math. **2012**, 541348 (2012)
2. Arnold, V.I., Afraimovich, V.S., Il'yashenko, Yu.S., Shil'niko, L.P.: Theory of bifurcations. In: Arnold, V. (ed.) Dynamical Systems, vol. 5. Encyclopedia of Mathematical Sciences. Springer, New York (1994)
3. Babushok, V.I., Goldshtein, V.M., Sobolev, V.A.: Critical condition for the thermal explosion with reactant consumption. Combust. Sci. Tech. **70**, 81–89 (1990)
4. Baris, Ya.S.: Integral manifolds of irregularly perturbed differential systems. Ukr. Math. J. **20**(4), 379–387 (1968)
5. Baris, Ya.S., Fodchuk, V.I.: Investigation of bounded solutions of nonlinear irregularly perturbed systems by the integral manifold method. Ukr. Math. J. **22**(1), 1–8 (1970)
6. Bellman, R.: Perturbation Techniques in Mathematics, Physics and Engineering. Holt, Rinehart and Winston, New York (1964)
7. Benoit, E.: Systèmes lents-rapides dans R^3 et leurs canards. Soc. Math. Fr. Astérisque **109–110**, 159–191 (1983)
8. Benoit, E., Callot, J.L., Diener, F., Diener M.: Chasse au canard. Collect. Math. **31–32**(1–3), 37–119 (1981–1982)
9. Bender, C.M., Orszag, S.A.: Advanced Mathematical Methods for Scientists and Engineers. I. Asymptotic Methods and Perturbation Theory. Springer, New York (1999)
10. Bobkova, A.S.: Duck trajectories in multidimensional singularly perturbed systems with a single fast variable. Differ. Equ. **40**(10), 1373–1382 (2004)
11. Bobkova, A.S., Kolesov, A.Yu., Rozov, N.Kh.: The "duck survival" problem in three-dimensional singularly perturbed systems with two slow variables. Math. Notes **71**(5–6), 749–760 (2002)
12. Bogatyrev, S.V., Sobolev, V.A.: Separation of rapid and slow motions in problems of the dynamics of systems of rigid bodies and gyroscopes. J. Appl. Math. Mech. **52**(1), 34–41 (1988)
13. Bogolyubov, N.N.: On Some Statistical Methods in Mathematical Physics (in Russian). Izd-vo Ukrainian Akademii Nauk, Kiev (1945)
14. Bogolyubov, N.N., Mitropolsky, Yu.A.: Asymptotic Methods in the Theory of Nonlinear Oscillations. Gordon and Breach, New York (1961)
15. Bogolyubov, N.N., Mitropolsky, Yu.A.: The Method of Integral Manifolds in Nonlinear Mechanics. Contributions to Differential Equations, vol. 2, pp. 123–196. Wiley, New York (1963)

16. Borovskikh, A.V.: Investigation of relaxation oscillations with the use of constructive nonstandard analysis: II. Differ. Equ. **40**(4), 491–501 (2004)
17. Boyce, W.E., DiPrima, R.C.: Elementary Differential Equations and Boundary Value Problems. Wiley, Hoboken (2008)
18. Brøns, M.: Bifurcations and instabilities in the Greitzer model for compressor system surge. Math. Eng. Ind. **2**, 51–63 (1988)
19. Brøns, M.: Relaxation oscillations and canards in a nonlinear model of discontinuous plastic deformation in metals at very low temperatures. Proc. R. Soc. A **461**, 2289–2302 (2005)
20. Brøns, M., Bar–Eli, K.: Canard explosion and excitation in a model of the Belousov–Zhabotinsky reaction. J. Phys. Chem. **95**, 8706–8713 (1991)
21. Brøns, M., Bar–Eli, K.: Asymptotic analysis of canards in the EOE equations and the role of the inflection line. Proc. Lond. R. Soc. Ser. A. **445**, 305–322 (1994)
22. Brøns, M., Kaasen, R.: Canards and mixed–mode oscillations in a forest pest model. Theor. Popul. Biol. **77**(4), 238–242 (2010)
23. Brøns, M., Sturis, J.: Explosion of limit cycles and chaotic waves in a simple nonlinear chemical system. Phys. Rev. E **64**, 026209 (2001)
24. Buric, L., Klic, A., Purmova, L.: Canard solutions and travelling waves in the spruce budworm population model. Appl. Math. Comput. **183**(2), 1039–1051 (2006)
25. Bush, A.W.: Perturbation Methods for Engineers and Scientists. CRC Press, Boca Raton (1992)
26. Butuzov, V.F., Nefedov, N.N., Schneider, K.R.: Singularly perturbed boundary value problems in case of exchange of stabilities. J. Math. Sci. **121**(1), 1973–2079 (2004)
27. Carr, J.: Applications of Centre Manifold Theory. Springer, New York (1981)
28. Cole, J.D.: Perturbation Methods in Applied Mathematics. Blaisdell Publishing Company, Waltham (1968)
29. Daletskii, Yu.L., Krein, M.G.: Stability of Solutions of Differential Equations in Banach Space. AMS, New York (1974)
30. De Maesschalck, P., Dumortier, F.: Canard solutions at non-generic turning points. Trans. AMS **358**(5), 2291–2334 (2006)
31. Desroches, M., Jeffrey, M.R.: Canard and curvature: the "smallness of ε" in slow–fast dynamics. Proc. R. Soc. A. **467**(2132), 2404–2421 (2011)
32. Desroches, M., Krauskopf, B., Osinga, H.M.: Numerical continuation of canard orbits in slow–fast dynamical systems. Nonlinearity **23**(3), 739–765 (2010)
33. Desroches, M., Guckenheimer, J., Krauskopf, B., Kuehn, C., Osinga, H.M., Wechselberge M.: Mixed–mode oscillations with multiple time scales. SIAM Rev. **54**(2), 211–288 (2012)
34. Desroches, M., Freire, E., Ponce, E., Hogan, Jh.S., Thota, P.: Canards in piecewise-linear systems: explosions and super-explosions. Proc. R. Soc. A. **469**(2154), 20120603 (2013)
35. Diener, M.: Nessie et Les Canards. Publication IRMA, Strasbourg (1979)
36. Diener, F., Diener, M.: Nonstandard Analysis in Practice. Springer, Berlin/New York (1995)
37. Dombrovskii, L.A., Zaichik, L.I.: Conditions of thermal explosion in a radiating gas with polydisperse liquid fuel. High Temp. **39**(4), 604–611 (2001)
38. Duan, J., Potzsche, C., Siegmund, S.: Slow integral manifolds for Lagrangian fluid dynamics in unsteady geophysical flows. Physica D **233**, 73–82 (2007)
39. Dumortier, F., Roussarie, R.: Canard cycles and center manifolds. Mem. Am. Math. Soc. **121**(577) (1996)
40. Dumortier, F., Roussarie, R.: Geometric singular perturbation theory beyond normal hyperbolicity. In: Jones, C.K.R.T., Khibnik A. (eds.) Multiple–Time–Scale Dynamical Systems. Volumes in Mathematics and its Applications, vol. 122, pp. 29–63. Springer, New York (2001)
41. Eckhaus, M.W.: Asymptotic Analysis of Singular Perturbations. Studies in Mathematics and Its Applications, vol. 9. North-Holland, Amsterdam/New York (1979)
42. Eckhaus, M.W.: Relaxation oscillations including a standard chase on French ducks. In: Verhulst. F. (ed.) Asymptotic Analysis II — Surveys and New Trends. Lecture Notes in Mathematics, vol. 985, pp. 449–494. Springer, Berlin (1983)

43. El-Sayed, S.A.: Adiabatic thermal explosion of a gas–solid mixture. Combust. Sci. Technol. **176**(2), 237–256 (2004)

44. Fehrst, A.: Enzyme Structure and Mechanisms. W.F. Freeman, New York (1985)

45. Fenichel, N.: Geometric singular perturbation theory for ordinary differential equations. J. Diff. Eq. **31**, 53–98 (1979)

46. Fraser, S.J.: The steady state and equilibrium approximations: a geometric picture. J. Chem. Phys. **88**, 4732–4738 (1988)

47. Freire, E., Gamero, E., Rodriguez-Luis, A.J.: First-order approximation for canard periodic orbits in a van der Pol electronic oscillator. Appl. Math. Lett. **12**, 73–78 (1999)

48. Fridman, E.: Asymptotics of integral manifolds and decomposition of singularly perturbed systems of neutral type. Differ. Equ. **26**, 457–467 (1990)

49. Fridman, E.: Decomposition of linear optimal singularly perturbed systems with time delay. Autom. Remote Control **51**, 1518–1527 (1990)

50. Fridman, E.: Decomposition of boundary problems for singularly perturbed systems of neutral type in conditionally stable case. Differ. Equ. **28**(6), 800–810 (1992)

51. Fruchard, A., Schafke, R.: Composite Asymptotic Expansions. Lecture Notes in Mathematics, vol. 2066. Springer, Heidelberg/New York (2013)

52. Ghanes, M., Hilairet, M., Barbot, J.P., Bethoux, O.: Singular perturbation control for coordination of converters in a fuel cell system. Electrimacs, Cergy-Pontoise, June 2011

53. Ghorbel, F., Spong, M.W.: Integral manifolds of singularly perturbed systems with application to rigid-link flexible-joint multibody systems. Int. J. Non Linear Mech. **35**, 133–155 (2000)

54. Gol'dshtein, V.M., Sobolev, V.A.: A Qualitative Analysis of Singularly Perturbed Systems (in Russian). Institut matematiki SO AN SSSR, Novosibirsk (1988)

55. Gol'dshtein, V. M., Sobolev, V.A.: Integral manifolds in chemical kinetics and combustion. In: Singularity Theory and Some Problems of Functional Analysis. AMS Translations. Series 2, vol. 153, pp. 73–92. AMS, Providence (1992)

56. Gol'dshtein, V., Zinoviev, A., Sobolev, V., Shchepakina, E.: Criterion for thermal explosion with reactant consumption in a dusty gas. Proc. Lond. R. Soc. Ser. A. **452**, 2103–2119 (1996)

57. Golodova, E.S., Shchepakina, E.A.: Modeling of safe combustion at the maximum temperature. Math. Models Comput. Simul. **1**(2), 322–334 (2009)

58. Gorban, A.N., Karlin, I.V.: Invariant Manifolds for Physical and Chemical Kinetics. Lecture Notes in Physics, vol. 660. Springer, Berlin/Heidelberg (2005)

59. Gorelov, G.N., Sobolev, V.A.: Mathematical modeling of critical phenomena in thermal explosion theory. Combust. Flame **87**, 203–210 (1991)

60. Gorelov, G.N., Sobolev, V.A.: Duck–trajectories in a thermal explosion problem. Appl. Math. Lett. **5**(6), 3–6 (1992)

61. Gorelov, G.N., Shchepakina, E.A., Sobolev V.A.: Canards and critical behavior in autocatalytic combustion models. J. Eng. Math. **56**, 143–160 (2006)

62. Grasman, J.: Asymptotic Methods for Relaxation Oscillations and Applications. Springer, New York (1987)

63. Grasman, J.: Relaxation Oscillations. Encyclopedia of Complexity and Systems Science, pp. 7602–7616. Springer, New York (2009)

64. Grasman, J., Wentzel, J.J.: Co-existence of a limit cycle and an equilibrium in Kaldor's business cycle model and its consequences. J. Econ. Behav. Organ. **01**, 369–377 (1994)

65. Gray, B.F.: Critical behaviour in chemical reacting systems: 2. An exactly soluble model. Combust. Flame **21**, 317–325 (1973)

66. Gu, Z., Nefedov, N.N., O'Malley, R.E.: On singular singularly perturbed initial values problems. SIAM J. Appl. Math. **49**, 1–25 (1989)

67. Guckenheimer, J., Towards a global theory of singularly perturbed systems. Prog. Nonlinear Diff. Eqns. Appl. **19**, 214–225 (1996)

68. Guckenheimer, J., Johnson, T., Meerkamp, P.: Rigorous enclosures of a slow manifold. SIAM J. Appl. Dyn. Syst. **11**(3), 831–863 (2012)

69. Guckenheimer, J., Holmes, P.: Nonlinear Oscillations, Dynamical Systems and Bifurcation of Vector Fields. Springer, New York (1983)

70. Hadamard, J.: Sur linteration et les solutions asymptotiques des equations differentielles. Bull. Soc. Math. Fr. **29**, 224–228 (1991)
71. Hale, J.: Integral manifolds of perturbed differential systems. Ann. Math. **73**(3), 496–531 (1961)
72. Hale, J.: Oscillations in Nonlinear Systems. Dover, New York (1963)
73. Hale, J.: Ordinary differential equations. Wiley, New York (1969)
74. Hausrath, A.R.: Stability in the critical case of purely imaginary roots for neutral functional differential equations. J. Differ. Equ. **13** 329–357 (1973)
75. Henry, D.: Geometrical Theory of Semilinear Parabolic Equations. Lecture Notes in Mathematics, vol. 804. Springer, New York/Berlin/Heidelberg (1981)
76. Hinch, E.J.: Perturbation Methods. Cambridge University Press, Cambridge (1991)
77. Holmes, M.H.: Introduction to Perturbation Methods. Springer, New York (1995)
78. Huyet, G., Porta, P.A., Hegarty, S.P., McInerney, J.G., Holland, F.: A low-dimensional dynamical system to describe low-frequency fluctuations in a semiconductor laser with optical feedback. Opt. Commun. **180**, 339–344 (2000)
79. Huzak, R., De Maesschalck, P., Dumortier, F.: Limit cycles in slow–fast codimension 3 saddle and elliptic bifurcations. J. Differ. Equ. **255**(11), 4012–4051 (2003)
80. Johnson, R.S.: Singular Perturbation Theory. Springer, New York (2005)
81. Jones, C.K.R.T.: Geometric Singular Perturbation Theory. Lecture Notes in Mathematics, vol. 1609, pp. 44–118. Springer, Berlin/Heidelberg (1995)
82. Kakiuchi, N., Tchizawa, K.: On an explicit duck solution and delay in the Fitzhugh–Nagumo equation. J. Diff. Equ. **141**, 327–339 (1997)
83. Kalachev, L.V., O'Malley, R.E.: The regularization of linear differential-algebraic equations. SIAM J. Math. Anal. **27**(1), 258–273 (1996)
84. Kaper, H.G., Kaper, T.J.: Asymptotic analysis of two reduction methods for systems of chemical reactions. Physica D **165**, 66–93 (2002)
85. Keener, J.P.: Principles of Applied Mathematics: Transformation and Approximation. Perseus Books, Cambridge (2000)
86. Kevorkian, J.: The Two Variable Expansion Procedure for the Approximate Solution of Certain Non-linear Differential Equations. Lectures in Applied Mathematics, vol. 7: Space Mathematics, part. III, pp. 206–275. American Mathematical Society, Providence (1966)
87. Kevorkian, J., Cole, J.D.: Perturbation Methods in Applied Mathematics. Springer, Berlin (1990)
88. Kevorkian, J., Cole, J.D.: Multiple Scales and Singular Perturbation Methods. Springer, New York (1996)
89. Kitaeva, E., Sobolev, V.: Numerical determination of bounded solutions to discrete singularly perturbed equations and critical combustion regimes. Comput. Math. Math. Phys. **45**(1), 52–82 (2005)
90. Knobloch, H.W., Aulback, B.: Singular perturbations and integral manifolds. J. Math. Phys. Sci. **18**(5), 415–424 (1984)
91. Kobrin, A.I., Martynenko, Yu.G.: Application of the theory of singularly perturbed equations to the study of gyroscopic systems (in Russian). Sov. Phys. Dokl. **21**(7–12), 498–500 (1976)
92. Kokotović, P.V., Khalil, K.H., O'Reilly, J.: Singular Perturbation Methods in Control: Analysis and Design. SIAM, Philadelphia (1986)
93. Kononenko, L.I., Sobolev, V.A.: Asymptotic expansion of slow integral manifolds. Sib. Math. J. **35**, 1119–1132 (1994)
94. Koshlyakov, V.N., Sobolev, V.A.: Permissibility of use of precessional equations of gyroscopic compasses. Mech. Solids **4**, 17–22 (1998)
95. Lagerstrom, P.A.: Matched Asymptotic Expansions: Ideas and Techniques. Springer, New York (2010)
96. Lam, S.H., Goussis, D.M.: The GSP method for simplifying kinetics. Int. J. Chem. Kinet. **26**, 461–486 (1994)
97. Lang, R., Kobayashi, K.: External optical feedback effects on semiconductor injection laser properties. IEEE J. Quantum Electron. **16**, 347–355 (1980)

98. Lancaster, P.: Theory of Matrices. Academic, New York/London (1969)
99. Lomov, S.: Introduction to the General Theory of Singular Perturbations. AMS Translations of Mathematical Monographs Series, vol. 112. AMS, Providence (1992)
100. Lyapunov, A.M.: The General Problem of the Stability of Motion (in Russian), Doctoral dissertation, Univ. Kharkov (1892). Liapounoff, A.: Probleme general de la stabilite du mouvement, Annales de la faculte des sciences de Toulouse, Ser. 2. **9**, 203–474 (1907). English translations: Stability of Motion, Academic, New York/London (1966)
101. Maas, U., Pope, S.B.: Simplifying chemical kinetics: Intrinsic low-dimensional manifolds in composition space. Combust. Flame **88**, 239–264 (1992)
102. Magnus, K.: The Gyroscope. Göttingen, Phywe (1983)
103. Mandel, P., Erneux T.: The slow passage through a steady bifurcation: delay and memory effects. J. Stat. Phys. **48**(5–6), 1059–1070 (1987)
104. Marino, F., Catalán, G., Sánchez, P., Balle, S., Piro, O.: Thermo–optical "canard orbits" and excitable limit cycles. Phys. Rev. Lett. **92**, 073901 (2004)
105. Martins, J.A.C., Rebrova, N., Sobolev, V.A.: On the (in)stability of quasi-static paths of smooth systems: definitions and sufficient conditions. Math. Methods Appl. Sci. **29**(6), 741–750 (2006)
106. Matyukhin, V.I.: Stability of the manifolds of controlled motions of a manipulator. Autom. Remote Control **59**(4), 494–501 (1998)
107. McIntosh, A.C.: Semenov approach to the modelling of thermal runawway of damp combustible material. IMA J. Appl. Math. **51**, 217–237 (1993)
108. McIntosh, A.C., Gray, B.F., Wake, G.C.: Analysis of the bifurcational behaviour of a simple model of vapour ignition in porous material. Proc. R. Soc. Lond. A. **453**(1957), 281–301 (1997)
109. Merkin, D.R.: Introduction to the Theory of Stability. Texts in Applied Mathematics, vol. 24. Springer, New-York (1996)
110. Mishchenko, E.F., Kolesov, Yu.S., Kolesov, A.Yu., Rozov, N.Kh.: Asymptotic Methods in Singularly Perturbed Systems. Plenum Press, New York (1995)
111. Mishchenko, E.F., Rozov, N.Kh.: Differential Equations with Small Parameters and Relaxation Oscillations. Plenum Press, New York (1980)
112. Mitropol'skii, Yu.: The method of integral manifolds in the theory of nonlinear oscillations. In: International Symposium on Nonlinear Differential Equations and Non-Linear Mechanics, pp. 1–15. Academic, New York/London (1963)
113. Mitropol'skii, Yu.A., Lykova, O.B.: Lectures on the Method of Integral Manifolds (in Russian). Izd. Naukova Dumka, Kiev (1968)
114. Mitropol'skii, Yu.A., Lykova, O.B.: Integral Manifolds in Nonlinear Mechanics (in Russian, MR 51:1025). Nauka, Moscow (1975)
115. Moehlis, J.: Canards in a surface oxidation reaction. J. Non Linear Sci. **12**, 319–345 (2002)
116. Mograbi, E., Bar-Ziv, E.: On the asymptotic solution of the Maxey–Riley equation. Phys. Fluids **18**(5), 051704 (2006)
117. Mortell, M.P., O'Malley, R.E., Pokrovskii, A., Sobolev, V.A. (eds.): Singular Perturbation and Hysteresis. SIAM, Philadelphia (2005)
118. Murray, J.D.: Lectures on Nonlinear Differential Equation Models in Biology. Clarendon Press, Oxford (1977)
119. Murray, J.D.: Mathematical Biology I. An Introduction. Springer, New York (2001)
120. Naidu, D.S.: Singular perturbations and time scales in control theory and applications: an overview. Dyn. Continuous Discrete Impulsive Syst. Ser. B Appl. Algorithms **9**(2), 233–278 (2002)
121. Nayfeh, A.H.: Perturbations Methods. Wiley, New York (1973)
122. Nayfeh, A.H.: Introduction to Perturbation Techniques. Wiley, New York (1995)
123. Nefedov, N.N., Schneider, K.R.: On immediate–delayed exchange of stabilities and periodic forced canards. Comput. Math. Math. Phys. **48**(1), 43–58 (2008). doi:10.1134/S0965542508010041 (2012)

124. Neishtadt, A.I.: On delayed stability loss under dynamical bifurcation. I. Differ. Equ. **23**, 2060–2067 (1987)
125. Neishtadt, A.I.: On delayed stability loss under dynamical bifurcation. II. Differ. Equ. **24**, 226–233 (1988)
126. Nipp, K., Stoffer, D.: Invariant Manifolds in Discrete and Continuous Dynamical Systems. EMS Tracts in Mathematics, vol. 21. European Mathematical Society Publishing House, Zürich (2013)
127. O'Malley, R.E.: Introduction to Singular Perturbations. Academic, New York (1974)
128. O'Malley, R.E.: High gain feedback systems as singular singular–perturbation problems. In: Prepr. techn. pap., 18th Joint Automat. Control Conf., New York, pp. 1278–1281 (1977)
129. O'Malley, R.E.: Singular Perturbations and Optimal Control. Springer Lecture Notes in Mathematics, vol. 680. Springer, New York (1978)
130. O'Malley, R.E.: Singular perturbation methods for ordinary differential equations. Applied Mathematical Sciences, vol. 89. Springer, New York (1991)
131. Osintsev, M.S., Sobolev, V.A.: Dimensionality reduction in optimal control and estimation problems for systems of solid bodies with low dissipation. Autom. Remote Control **74**(8), 1334–1347 (2013)
132. Osintsev, M.S., Sobolev, V.A.: Reduction of dimension of optimal estimation problems for dynamical systems with singular perturbations. Comput. Math. Math. Phys. **54**(1), 45–58 (2014)
133. Pachpatte, B.: Integral and Finite Difference Inequalities and Applications, Vol. 205. Elsevier, Oxford (2006)
134. Pendyukhova, N.V., Sobolev, V.A., Strygin, V.V.: Motion of rigid body with gyroscope and moving mass points. Mech. Solids **3**, 12–18 (1986)
135. Peng, B., Gáspár, V., Showalter., K.: False bifurcations in chemical systems: Canards. Philos. Trans. R. Soc. Lond. A. **337**, 275–289 (1991)
136. Perron, O.: Über Stabilität und asymptotisches Verhalten der Integrale von Differentialgleichungssystemen (in German). Math. Z. **29**(1), 129–160 (1929)
137. Petrov, V., Scott, S.K., Showalter, K.: Mixed–mode oscillations in chemical systems. J. Chem. Phys. **97**, 6191–6198 (1992)
138. Pliss, V.A.: A reduction principle in the theory of stability of motion (in Russian, MR 32:7861). Izv. Akad. Nauk SSSR Ser. Mat. **28**, 1297–1324 (1964)
139. Poincare, H.: Les Methodes nouvelles de la Mecanique Celeste. Tome I. Paris, Gauthier-Villars (1892)
140. Pokrovskii, A., Sobolev, V.: A naive view of time relaxation and hysteresis. In: Mortell, M., O'Malley, R., Pokrovskii, A., Sobolev, V. (eds.) Singular Perturbations and Hysteresis, pp. 1–60. SIAM, Philadelphia (2005)
141. Pokrovskii, A., Shchepakina E., Sobolev, V.: Canard doublet in a Lotka-Volterra type model. J. Phys. Conf. Ser. **138**, 012019 (2008)
142. Pokrovskii, A., Rachinskii, D., Sobolev, V., Zhezherun, A.: Topological degree in analysis of canard-type trajectories in 3-D systems. Appl. Anal. **90**(7), 1123–1139 (2011)
143. Popovic, N.: Mixed–mode dynamics and the canard phenomenon: towards a classification. J. Phys. Conf. Ser. **138**, 012020 (2008)
144. Prohens, R., Teruel, A.E.: Canard trajectories in 3d piecewise linear systems. Discrete Continuous Dyn. Syst. **33**(10), 4595–4611 (2013)
145. Roussel, M.R., Fraser, S.J.: Geometry of of the steady–state approximation: perturbation and accelerated convergence method. J. Chem. Phys. **93**, 1072–1081 (1990)
146. Rudin, W.: Principles of Mathematical Analysis. McGraw-Hill, New York (1976)
147. Sazhin, S.S., Shchepakina, E., Sobolev, V.: Order reduction of a non-Lipschitzian model of monodisperse spray ignition. Math. Comput. Model. **52**(3–4), 529–537 (2010)
148. Schneider, K.R., Wilhelm, T.: Model reduction by extended quasi-steady state assumption. J. Math. Biology. **40**, 443–450 (2000)
149. Schneider, K., Shchepakina, E., Sobolev, V.: A new type of travelling wave. Math. Methods Appl. Sci. **26**, 1349–1361 (2003)

150. Schneider, K.R., Shchetinina, E., Sobolev, V.A.: Control of integral manifolds loosing their attractivity in time. J. Math. Anal. Appl. **315**(2), 740–757 (2006)
151. Sekikawa, M., Inaba, N., Tsubouchi, T.: Chaos via duck solution breakdown in a piecewise linear van der Pol oscillator driven by an extremely small periodic perturbation. Physica D **194**, 227–249 (2004)
152. Semenov, N.N.: Zur theorie des verbrennungsprozesses (in German). Z. Phys. Chem. **48**, 571–581 (1928)
153. Simmonds, J.G., Mann, J.E. Jr.: A First Look at Perturbation Theory. Dover, New York (1998)
154. Shchepakina, E.: Critical conditions of self–ignition in dusty media. J. Adv. Chem. Phys. **20**(7), 3–9 (2001)
155. Shchepakina, E.: Slow integral manifolds with stability change in the case of a fast vector variable. Differ. Equ. **38**, 1146–1452 (2002)
156. Shchepakina, E.: Black swans and canards in a self-ignition problem. Non Linear Anal. Real World Appl. **4**, 45–50 (2003)
157. Shchepakina, E.A.: Two forms of stability change for integral manifolds. Differ. Equ. **40**(5), 766–769 (2004)
158. Shchepakina, E.: Canards and black swans in a model of a 3-D autocatalator. J. Phys. Conf. Ser. **22**, 194–207 (2005)
159. Shchepakina, E., Korotkova, O.: Condition for canard explosion in a semiconductor optical amplifier. J. Opt. Soc. Am. B. **28**, 1988–1993 (2011)
160. Shchepakina, E., Korotkova, O.: Canard explosion in chemical and optical systems. Discrete Continuous Dyn. Syst. **18**, 495–5512 (2013)
161. Shchepakina, E., Sobolev, V.: Standard chase on black swans and canards, Weierstraß–Institut für Angewandte Analysis und Stochastik. Preprint 426. WIAS, Berlin (1998)
162. Shchepakina, E., Sobolev, V.: Attracting/repelling invariant manifolds. Stab. Control Theory Appl. **3**(3), 263–274 (2000)
163. Shchepakina, E., Sobolev, V.: Integral manifolds, canards and black swans. Non Linear Anal. Ser. A Theory Methods **44**(7), 897–908 (2001)
164. Shchepakina, E., Sobolev, V.: Black swans and canards in laser and combustion models. In: Mortell, M., O'Malley, R., Pokrovskii, A., Sobolev, V. (eds.) Singular Perturbations and Hysteresis, pp. 207–256. SIAM, Philadelphia (2005)
165. Shchepakina, E.A., Sobolev, V.A.: Modelling of critical phenomena for ignition of metal particles. J. Phys. Conf. Ser. **138**, 012025 (2008)
166. Shouman, A.R.: Solution to the dusty gas explosion problem with reactant consumption part I: the adiabatic case. Combust. Flame **119**(1), 189–194 (1999)
167. Smetannikova, E., Sobolev, V.: Periodic singularly perturbed problem for the matrix Riccati equation. Differ. Equ. **41**(4), 529–537 (2005)
168. Smetannikova, E., Sobolev, V.: Regularization of cheap periodic control problems. Autom. Remote Control **66**(6), 903–916 (2005)
169. Smith, D.R.: Singular–Perturbation Theory: An Introduction with Applications. Cambridge University Press, Cambridge (1985)
170. Sobolev, V.A.: Integral manifolds and decomposition of singularly perturbed systems. Syst. Control Lett. **5**, 169–179 (1984)
171. Sobolev, V.A.: Decomposition of control systems with singular perturbations. In: Proceedings of the 10th Congress of IFAC, Munich, vol. 8, pp. 172–176 (1987)
172. Sobolev, V.A.: Decomposition of linear singularly perturbed systems. Acta Math. Hung. **49**, 365–376 (1987)
173. Sobolev, V.A.: Integral manifolds, singular perturbations and optimal control. Ukr. Math. J. **39**, 95–99 (1987)
174. Sobolev, V.A.: Integral manifolds, stability and decomposition of singularly perturbed systems in Banach space. Acta Sci. Math. **51**(3–4), 491–500 (1987)
175. Sobolev, V.A.: Integral manifolds and decomposition of nonlinear differential systems. Stud. Sci. Math. **23**, 73–79 (1988)

176. Sobolev, V.A.: Geometrical theory of singularly perturbed control systems. In: Proceedings of the 11th Congress of IFAC, Tallinn, vol. 6, pp. 163–168 (1990)
177. Sobolev, V.A.: Nonlocal integral manifolds and decoupling of nonlinear parabolic systems. In: Global Analysis — Studies and Applications IV. Springer Lecture Notes in Mathematics, vol. 1453, pp. 101–108. Springer, Berlin/Heidelberg (1990)
178. Sobolev, V.A.: Singular perturbations in linearly quadratic optimal control problems. Autom. Remote Control. **52**(2), 180–189 (1991)
179. Sobolev, V.: Black swans and canards in laser and combustion models. In: Mortell, M., O'Malley, R., Pokrovskii, A., Sobolev, V. (eds.) Singular Perturbations and Hysteresis, pp. 153–206. SIAM, Philadelphia (2005)
180. Sobolev, V.A.: Canard cascades. Discrete Continuous Dyn. Syst. **18**, 513–521 (2013)
181. Sobolev, V.A., Chernyshov, K.I.: Singularly perturbed differential equation with a Fredholm operator applied to the derivative. Differ. Equ. **25**(2), 181–190 (1989)
182. Sobolev, V.A., Fridman, L.M.: Decomposition of systems with discontinuous controls whose elements operate at different speeds. Autom. Remote Control **49**, 288–292 (1988)
183. Sobolev, V.A., Shchepakina, E.A.: Self–ignition of laden medium. J. Combust. Explos. Shock Waves **29**(3), 378–381 (1993)
184. Sobolev, V.A., Shchepakina, E.A.: Duck trajectories in a problem of combustion theory. Differ. Equ. **32**, 1177–1186 (1996)
185. Sobolev, V.A., Shchepakina, E.A.: Model Reduction and Critical Phenomena in Macrokinetics (in Russian). Moscow, Fizmatlit (2010)
186. Sobolev, V.A., Strygin, V.V.: Permissibility of changing over to precession equations of gyroscopic systems. Mech. Solids **5**, 7–13 (1978)
187. Sobolev, V.A., Tropkina, E.A.: Asymptotic expansions of slow invariant manifolds and reduction of chemical kinetics models. Comput. Math. Math. Phys. **52**, 75–89 (2012)
188. Sobolev, V.A., Andreev, I.A., Shchepakina, E.A.: Modeling of critical phenomena in autocatalytic burning problems. In: Proceedings of the 15th IMACS World Congress, Berlin, vol. 6, pp. 317–322 (1997)
189. Sobolev, V., Schneider, K., Shchepakina, E.: Three types of non-adiabatic combustion waves in the case of autocatalytic reaction. J. Adv. Chem. Phys. **24**(6), 63–69 (2005)
190. Spong, M.W.: Modeling and control of elastic joint robots. J. Dyn. Syst. Meas. Control **109**(4), 310–319 (1987)
191. Spong, M.W., Khorasani, K., Kokotovic, P.V.: An integral manifold approach to feedback control of flexible joint robots. IEEE J. Rob. Autom. **3**(4), 291–301 (1987)
192. Stewart, G.W.: Introduction to Matrix Computations. Academic, New York (1973)
193. Straughan, B.: Explosive Instabilities in Mechanics. Springer, Berlin/Heidelberg (1998)
194. Strogatz, S.H.: Nonlinear Dynamics and Chaos. Westview Press, Cambridge (2000)
195. Strygin, V.V., Sobolev, V.A.: Effect of geometric and kinetic parameters and energy dissipation on orientation stability of satellites with double spin. Cosmic Res. **14**(3), 331–335 (1976)
196. Strygin, V.V., Sobolev, V.A.: Asymptotic methods in the problem of stabilization of rotating bodies by using passive dampers. Mech. Solids **5**, 19–25 (1977)
197. Strygin, V.V., Sobolev, V.A.: Separation of Motions by the Integral Manifolds Method (in Russian, MR 89k:34071). Nauka, Moscow (1988)
198. Tchizawa, K.: On relative stability in 4-dimensional duck solutions. J. Math. Syst. Sci. **2**(9), 558–563 (2012)
199. Tikhonov, A.N.: Systems of differential equations with small parameters multiplying the derivatives. Matem. Sb. **31**(3), 575–586 (1952)
200. Tsengand, H.C., Siljak, D.D.: A learning scheme for dynamic neural networks: equilibrium manifold and connective stability. Neural Netw. **8**(6), 853–864 (1995)
201. Utkin, V.I.: Application of equivalent control method to the systems with large feedback gain. IEEE Trans. Automat. Control **23**(3), 484–486 (1977)
202. Utkin, V.I.: Sliding Modes and Their Applications in Variable Structure Systems. Mir, Moscow (1978)
203. Van Dyke, M.: Perturbation Methods in Fluid Dynamics. Parabolic Press, Stanford (1975)

204. Vasil'eva, A.B., Butuzov, V.F.: Singularly perturbed equations in the critical case. In: Tech. Report MRC–TSR 2039. University of Wisconsin, Madison (1980)
205. Vasil'eva, A.B., Butuzov, V.F., Kalachev, L.V.: The boundary function method for singular perturbation problems. In: Studies in Applied Mathematics, vol. 14. SIAM, Philadelphia (1995)
206. Vidilina, O.: Singular perturbations in time-optimal control problem. Stability Control Theory Appl. 6(1), 1–9 (2004)
207. Voropaeva, N.V.: Decomposition of problems of optimal control and estimation for discrete systems with fast and slow variables. Autom. Remote Control 69(6), 920–928 (2008)
208. Voropaeva, N.V., Sobolev, V.A.: A constructive method of decomposition of singularly perturbed nonlinear differential systems. Differ. Equ. 31(4), 528–537 (1995)
209. Voropaeva, N.V., Sobolev, V.A.: Decomposition of a linear–quadratic optimal control problem with fast and slow variables. Autom. Remote Control. 67(8), 1185–1193 (2006)
210. Voropaeva, N.V., Sobolev, V.A.: Geometrical Decomposition of Singularly Perturbed Systems (in Russian). Moscow, Fizmatlit (2009)
211. Wasow, W.: Asymptotic Expansions for Ordinary Differential Equations. Wiley, New York (1965)
212. Xie, F., Han, M.A.: Existence of canards under non-generic conditions. Chin. Ann. Math. Ser. B 30(3), 239–250 (2009)
213. Xie, F., Han, M.A., Zhang, W.J.: Canard phenomena in oscillations of a surface oxidation reaction. J. Non Linear Sci. 15(6), 363–386 (2005)
214. Xie, F., Han, M.A., Zhang, W.J.: Existence of canard manifolds in a class of singularly perturbed systems. Non Linear Anal. Theory Methods Appl. 64(3), 457–470 (2006)
215. Xie, F., Han, M.A., Zhang, W.J.: The persistence of canards in 3-D singularly perturbed systems with two fast variables. Asymptotic Anal. 47(1–2), 95–106 (2006)
216. Young, K.-K.D., Kokotović, P.V., Utkin, V.I.: A singular perturbation analysis of high-gain feedback systems. IEEE Trans. Automat. Control 22(6), 931–938 (1977)
217. Zadiraka, K.V.: On the integral manifold of a system of differential equations containing a small parameter (in Russian, MR 19:858a). Dokl. Akad. Nauk SSSR 115, 646–649 (1957)
218. Zadiraka, K.V.: On a non-local integral manifold of a singularly perturbed differential system (in Russian, MR 34:4633). Ukr. Math. Z. 17(1), 47–63 (1965) [Translated in AMS Transl., Ser. 2. 89, 29–49 (1970)]
219. Zeldovich, Ya.B., Barenblatt, G.I., Librovich, V.B., Makhviladze, G.M.: The Mathematical Theory of Combustion and Explosions. Consultants Bureau, New York (1985)
220. Zhou, Zh., Shen, J.: Delayed phenomenon of loss of stability of solutions in a second-order quasi-linear singularly perturbed boundary value problem with a turning point. Boundary Value Probl. 35 (2011). doi:10.1186/1687-2770-2011-35
221. Zvonkin, A.K., Shubin, M.A.: Non-standard analysis and singular perturbations of ordinary differential equations. Russ. Math. Surv. 39, 69–131 (1984)

Index

© Springer International Publishing Switzerland 2014
E. Shchepakina et al., *Singular Perturbations*, Lecture Notes in Mathematics 2114,
DOI 10.1007/978-3-319-09570-7

LECTURE NOTES IN MATHEMATICS

Edited by J.-M. Morel, B. Teissier; P.K. Maini

Editorial Policy (for the publication of monographs)

1. Lecture Notes aim to report new developments in all areas of mathematics and their applications - quickly, informally and at a high level. Mathematical texts analysing new developments in modelling and numerical simulation are welcome.

   Monograph manuscripts should be reasonably self-contained and rounded off. Thus they may, and often will, present not only results of the author but also related work by other people. They may be based on specialised lecture courses. Furthermore, the manuscripts should provide sufficient motivation, examples and applications. This clearly distinguishes Lecture Notes from journal articles or technical reports which normally are very concise. Articles intended for a journal but too long to be accepted by most journals, usually do not have this "lecture notes" character. For similar reasons it is unusual for doctoral theses to be accepted for the Lecture Notes series, though habilitation theses may be appropriate.

2. Manuscripts should be submitted either online at www.editorialmanager.com/lnm to Springer's mathematics editorial in Heidelberg, or to one of the series editors. In general, manuscripts will be sent out to 2 external referees for evaluation. If a decision cannot yet be reached on the basis of the first 2 reports, further referees may be contacted: The author will be informed of this. A final decision to publish can be made only on the basis of the complete manuscript, however a refereeing process leading to a preliminary decision can be based on a pre-final or incomplete manuscript. The strict minimum amount of material that will be considered should include a detailed outline describing the planned contents of each chapter, a bibliography and several sample chapters.

   Authors should be aware that incomplete or insufficiently close to final manuscripts almost always result in longer refereeing times and nevertheless unclear referees' recommendations, making further refereeing of a final draft necessary.

   Authors should also be aware that parallel submission of their manuscript to another publisher while under consideration for LNM will in general lead to immediate rejection.

3. Manuscripts should in general be submitted in English. Final manuscripts should contain at least 100 pages of mathematical text and should always include

   - a table of contents;
   - an informative introduction, with adequate motivation and perhaps some historical remarks: it should be accessible to a reader not intimately familiar with the topic treated;
   - a subject index: as a rule this is genuinely helpful for the reader.

   For evaluation purposes, manuscripts may be submitted in print or electronic form (print form is still preferred by most referees), in the latter case preferably as pdf- or zipped ps-files. Lecture Notes volumes are, as a rule, printed digitally from the authors' files. To ensure best results, authors are asked to use the LaTeX2e style files available from Springer's web-server at:

   ftp://ftp.springer.de/pub/tex/latex/svmonot1/ (for monographs) and
   ftp://ftp.springer.de/pub/tex/latex/svmultt1/ (for summer schools/tutorials).

Additional technical instructions, if necessary, are available on request from lnm@springer.com.

4. Careful preparation of the manuscripts will help keep production time short besides ensuring satisfactory appearance of the finished book in print and online. After acceptance of the manuscript authors will be asked to prepare the final LaTeX source files and also the corresponding dvi-, pdf- or zipped ps-file. The LaTeX source files are essential for producing the full-text online version of the book (see http://www.springerlink.com/openurl.asp?genre=journal&issn=0075-8434 for the existing online volumes of LNM). The actual production of a Lecture Notes volume takes approximately 12 weeks.

5. Authors receive a total of 50 free copies of their volume, but no royalties. They are entitled to a discount of 33.3 % on the price of Springer books purchased for their personal use, if ordering directly from Springer.

6. Commitment to publish is made by letter of intent rather than by signing a formal contract. Springer-Verlag secures the copyright for each volume. Authors are free to reuse material contained in their LNM volumes in later publications: a brief written (or e-mail) request for formal permission is sufficient.

Addresses:
Professor J.-M. Morel, CMLA,
École Normale Supérieure de Cachan,
61 Avenue du Président Wilson, 94235 Cachan Cedex, France
E-mail: morel@cmla.ens-cachan.fr

Professor B. Teissier, Institut Mathématique de Jussieu,
UMR 7586 du CNRS, Équipe "Géométrie et Dynamique",
175 rue du Chevaleret
75013 Paris, France
E-mail: teissier@math.jussieu.fr

For the "Mathematical Biosciences Subseries" of LNM:

Professor P. K. Maini, Center for Mathematical Biology,
Mathematical Institute, 24-29 St Giles,
Oxford OX1 3LP, UK
E-mail: maini@maths.ox.ac.uk

Springer, Mathematics Editorial, Tiergartenstr. 17,
69121 Heidelberg, Germany,
Tel.: +49 (6221) 4876-8259

Fax: +49 (6221) 4876-8259
E-mail: lnm@springer.com